BILLY MITCHELL

The Art of War

Series Editor, David T. Zabecki

BILLY MITCHELL

JAMES J. COOKE

LYNNE
RIENNER
PUBLISHERS

BOULDER
LONDON

Published in the United States of America in 2002 by
Lynne Rienner Publishers, Inc.
1800 30th Street, Boulder, Colorado 80301
www.rienner.com

and in the United Kingdom by
Lynne Rienner Publishers, Inc.
3 Henrietta Street, Covent Garden, London WC2E 8LU

Library of Congress Cataloging-in-Publication Data
Cooke, James J.
    Billy Mitchell / James J. Cooke.
      p. cm.
    Includes bibliographical references and index.
    ISBN 1-58826-082-8 (alk. paper)
    1. Mitchell, William, 1879–1936.   2. Generals—United States—Biography.
    3. Aeronautics, Military—United States—History—20th century.   4. United States.
    Army. Air Corps—History—20th century.   I. Title.
  UG626.2.M57 C66   2002
  358.4'0092—dc21

                                                     2002018899

British Cataloguing in Publication Data
A Cataloguing in Publication record for this book
is available from the British Library.

Printed and bound in the United States of America

  5   4   3   2   1

CONTENTS

v

PHOTOGRAPHS

PREFACE

A S A HISTORIAN AND A PROFESSOR OF HISTORY FOR THREE decades, I was aware that figures in history often take on a larger-than-life persona that sometimes cannot stand up under serious research. There is a school of historical research that delights in revealing the warts of great persons, and concludes that those bumps and blemishes demonstrate how these individuals really contributed little to history and are unworthy of modern respect. For example, some founding fathers of the United States, because they owned slaves, are now viewed as hypocrites who did not mean what they wrote in the Declaration of Independence or the Constitution. The student newspaper at the University of Mississippi, where I taught, ran an editorial excoriating every president from Abraham Lincoln on for their personal failings. Much of the information in the editorial was incorrect, but in the writers' minds they were in the mainstream of current thought.

There are dangers in writing a biography of a historical figure when he has been perceived as a "prophet without honor." It is doubly difficult to delve into a person's life and show that he was not all that he was thought to be, when his dramatic court-martial was depicted on the movie screen by such an icon as Gary Cooper. The problem is compounded when you consider that Billy Mitchell remains a hero of the United States Air Force, a military institution worthy of respect and support. When one flies into Milwaukee, Wisconsin, the plane lands at the airport named after Mitchell. In the terminal there is a fine Mitchell Gallery of Flight Air Museum, an impressive presentation that gladdens the heart of any historian because average folk who merely survived the mind-numbing vagaries of high school U.S. history spend a good deal of time looking at and learning from the well-presented displays.

My interest in Billy Mitchell began when I spent an academic year as a visiting professor at the Air War College at Maxwell Air Force Base in Montgomery, Alabama. At that time I was just starting research on my second book on World War I, *The U.S. Air Service in the Great War* (1996), and it would have been impossible to study the efforts of the United States in the air and not take a long look at the contributions of Billy Mitchell. Besides researching the documents and materials in the U.S. Air Force Historical Agency's archives, I read the printed diary of Colonel Frank P. Lahm, which indicated a deep-seated hostility toward Billy Mitchell felt by many in the American Expeditionary Forces (AEF) during the so-called Great War. This hostility was confirmed while I was doing research at the U.S. Army Military History Institute's archives at Carlisle Barracks, Pennsylvania, as I reviewed the papers of generals Hugh A. Drum and Dennis Nolan, both of whom served with Mitchell in the AEF and played major roles in subsequent events in Mitchell's life. I had seen the 1955 movie *The Court Martial of Billy Mitchell* several times, and I shared the popular view that if only the U.S. leadership of the late 1920s and 1930s had listened to Mitchell we would not have been so unprepared for war in 1941—and that possibly the Japanese would not have been able to carry out their surprise attack at Pearl Harbor on December 7. But the more research I did on World War I, the more I became convinced that a good biography of Billy Mitchell was needed; what was missing from previous works on Billy Mitchell was the man himself. Having myself endured the stresses and strains of war, I assumed that part of the antagonism toward Mitchell resulted from the natural clash of egos, perceptions, and roles in wartime; but there was something that ran deeper than just the normal competition between highly motivated, goal-oriented officers. Suppositions such as "If only America had listened to Mitchell" and "If only Congress had stepped in during the 1925 court-martial" are not history; they are flights of fancy. There were too many other factors at play that brought about the devastating circumstances of December 1941. Having said that, it is wise to point out that Mitchell was correct in his assessment of defenses in the Pacific area.

Colonel William Mitchell was a complex man—certainly not the Gary Cooperesque saint of the 1955 movie, nor the hated demon that some in the War Department believed him to be. He massed over 1,400 aircraft to support the U.S. attack at St. Mihiel in 1918, and he showed in 1921 that aircraft could sink a modern battleship. His devotion to airpower was sincere, and he firmly believed that air forces should be separate from the army and the navy. Mitchell was on target when he described the state of U.S. defenses in the Hawaiian Islands and the Philippines in the 1920s. He was brave in battle and inspired great loyalty in subordinates. Mitchell was a prolific author, having published three books on airpower and a vast number of articles in popular magazines. His ideas, rehashed many times, were available to the public, and he appeared to be closer to them than to the high-collared, grim-faced generals of the War Department. But there was another side to Mitchell. How much did he do for himself, and how much did he do "for the good of the service"? Mitchell was brash, contemptuous of superiors, unwilling to work within the army system, incapable of giving credit to others, and at times he could be unfeeling when his subordinates were in dire circumstances. He played at being rich—piling up debts, spending heavily, and borrowing money—and had one failed marriage. Finally, as a result of his actions, he lost the support of influential generals, including John J. Pershing, when they could have helped him the most. In 1925 the secretary of war described Mitchell's behavior as "lawless," as indeed it had become, and President Calvin Coolidge decided that the time had come to court-martial Billy Mitchell. Done right, a scholarly biography must look at all aspects of the materials available, make judgments, and overlook those who would idealize or demonize the subject. I have tried to do this while locating the person of Billy Mitchell and pointing out where he fits within the history of aviation in the United States.

To "find" Mitchell I had expert help from a number of first-rate archivists who made materials available and gave sound advice. Mitchell Yockelson of the Military Records of the National Archives first led me to Air Force archives and then, in their new facility at College Park, Maryland, to three dozen cartons that contain Mitchell's court-martial records and transcripts. No historian can do serious work without the archivist's expert advice and in-depth knowledge of the holdings. Yockelson and his colleagues, who have put up with me for five books now, certainly must represent the best in public service.

The staff of the Library of Congress assisted me greatly, making available their fine collections of personal papers. Ernest Emrich, Jeffery Flannery, Fred Bauman, and Bradley Gernand have an extensive knowledge of their holdings and, possibly of more importance, a deep love of American history and a sense of responsibility in preserving our nation's past. The most valuable of their offerings are the Mitchell papers, which contain vast numbers of letters and other documents. It seemed that Mitchell's mother never threw away a letter from her son, and particularly useful were those from the young lad, "Willie," while at boarding school at Racine College.

Every once in a while the historian stumbles upon a gem, and I did just that when I contacted the archives of the Golda Meir Library at the University of Wisconsin, Milwaukee. I was informed by a very fine archivist, Nicolas Weber, that the library holds the George Hardie Collection, which focuses on the history of aviation, and that a good part of the collection contains Billy Mitchell material. While researching this collection I discovered the library's Mitchell Family Papers, a small collection that turned out to provide vitally important insights. The Milwaukee County Historical Society and the Milwaukee Public Library were also most helpful in locating materials, especially newspaper accounts of Mitchell's father's spectacular divorce in the late 1870s.

Another valuable source of information was the Alumnae and Alumni Society of Vassar College in Poughkeepsie, New York. The executive director of the society, Terri O'Shea, provided a great deal of background material pertaining to Mitchell's first wife, Caroline Stoddard Mitchell. Billy Mitchell had four strong women in his life—his mother; his sister Harriet; Caroline Stoddard; and his second wife, Elizabeth Trumbull Miller. No biography of Mitchell would be complete without a good deal of attention paid to those women who did so much to shape his life from birth to death.

Special thanks must go to Roger G. Miller of the Air Force History Support Office at Bolling Air Force Base, Washington, D.C. Dr. Miller spent a vast amount of time in discussion and correspondence with me about Billy Mitchell and the Air Service. Without his guidance I might not have remained on track. I am also indebted to Christopher R. Paulson of the Racine Heritage of Racine, Wisconsin; Paul Woehrmann, Local History and Marine Collection of the Milwaukee Public Library; Phyllis Aurand of the Thomas Balch Library of Leesburg, Virginia; and Daniel Heath de Butts of the Fauquier Historical Society of Warrenton, Virginia. Robert Haws, chairman of the Department of History of the University of Mississippi, must be acknowledged for his support, as must T. J. Ray, professor emeritus of English, whose interest in aviation history made him an easy mark for listening to my ideas and, often, my problems. Wood Brown III, of Slidell and New Orleans, Louisiana, must be cited for his assistance in providing material on the story of the Montana Class battleships. Michelle Palmertree took many disks and converted them into readable pages because my computer skills are truly nonexistent. Equally bad is my use of commas, colons, semicolons, and the like. My wife of now forty years continues, after many books, articles, scholarly papers, and book reviews, to serve as my Editorial Queen. The family that suffers together through the nuances of English grammar stays together, or something like that.

— James J. Cooke

INTRODUCTION

I T WOULD BE VERY EASY TO BE MESMERIZED BY BILLY
Mitchell the prophet without honor in his time, the man who predicted
the Japanese attack on Pearl Harbor, the persecuted defender of the United
States he saw as poorly prepared for the next air war. For those who wish to
skim the surface of history the legendary Mitchell is comfortable, easy to
understand. If history is, however, the careful and accurate recording of
past events, then the comfortable is not always totally satisfactory. No person who has made a mark on the pages of history is one dimensional, and
to be sure Billy Mitchell made his mark. He is not a footnote to history; he
is an entire chapter, the subject of many books and articles. Many would
not want to believe there was a dark side to his personality. Billy Mitchell
was not the self-sacrificing airman who gave up everything at his court-
martial in 1925. He could be a shameless self-promoter, a difficult man to
deal with, a poor husband, and eventually a neglectful father. As a soldier
he could be a brilliant fighter but also a poor subordinate—a man great in
war, terrible in peace.

Despite his many short-comings, Mitchell earned his place in history
as a pioneer air war fighter who saw better than anyone else the great
potential of massed combat aircraft striking at an enemy on the battlefield.
He was not simply a visionary who wrote and talked: he assembled over
1,400 airplanes to pound the Germans at St. Mihiel in 1918. No one would
ever have doubted Mitchell's intelligence, and he was one of the first air
fighters to see the great potential of airpower in combined arms operations.

The World War I infantry division was a combined arms team, with
infantry and artillery working in concert with other elements of the division. Typically an aero-observation squadron and a balloon company with
one balloon was assigned to work with the division's artillery and the

1

infantry. The prevailing belief was that airpower was best used in support of and in concert with the ground combat forces; and given the technological limitations of the World War I battlefield, integration of air and ground went fairly well. Mitchell, however, went beyond the concept of air-ground cooperation and saw air as extending the battlefield, with bomber aircraft acting beyond the enemies' front lines, hitting supply routes, ammunition dumps, bridges, airfields, and German troop concentrations. Pursuit aircrafts' primary function was to gain air superiority, driving enemy reconnaissance and pursuit planes from the skies. Once that was done, pursuit aircraft could then strafe enemy positions (within the limits of safety) to facilitate the attacks of American infantry. Despite his mantle of Airpower Prophet, Mitchell remained fairly orthodox in his views of battle, looking primarily at the battle at hand. He diverged from the doctrines of the Great War mostly in his emerging concept of an independent air force, which eventually developed into his belief that massive air assets could by themselves affect the outcome of not only a battle, but a war.

As assistant to the chief of the Air Service, Billy Mitchell had many personal motives for his fight with the navy over the primacy of the battleship in 1921, but he did sink the great German warship *Ostfriesland* in a few minutes. Many other ships sank under the pounding of Mitchell's aerial bombardment, showing to all that great surface ships, once the pride of ocean-going navies, were vulnerable to attack by single-engine airplanes. Mitchell's crusade against the navy, however, clouded his vision and his usually fertile imagination regarding the potential of the aircraft carrier and the submarine. He did indeed see the grave dangers to American military power in the Pacific and in 1923 addressed them in a report of defenses on the Hawaiian Islands and the Philippine Islands. Knowing how aircraft could devastate military installations—especially those at Pearl Harbor in Hawaii and Clark Field in the Philippines—he wrote, with great accuracy, how such enemy attacks could be carried out. But his remedies for shortcomings of the defenses in the Pacific were far beyond the technological and financial capabilities of the United States at that time.

When war began in Europe in 1939, three years after Mitchell's death, the capabilities of the German Luftwaffe fighting in Poland and then in 1940 in the Low Countries and France motivated airpower preparedness advocates to resurrect Mitchell's ideas and warnings. Billy Mitchell was a highly literate man, and he wrote a massive number of books and articles for publication in civilian journals and magazines. Consequently, it was not difficult to find Mitchell's views, and many were reprinted in major popular magazines. As the smoke cleared over the ruins of the great American fleet at Pearl Harbor, Mitchell became even more important as a prophet who had been silenced by the powerful decisionmakers in Washington. Billy Mitchell's views had been repressed by those generals and politicians, or so presumes the conspiracy theory. Those who wanted to find sinister motives in the War Department overlooked the fact that many of the founding fathers of airpower, such as Mason Patrick, Benjamin Foulois, and James Fetchet, had reached similar conclusions but were able to state their opinions and positions without incurring the wrath of the president or those in position to affect the decisionmaking process.

After World War II Billy Mitchell was again cited as an authority when the question of the establishment of an independent air force was debated. Many of the great air generals of the Second World War, Henry "Hap" Arnold among them, had defended Mitchell at his famous 1925 court-martial. When the United States Air Force (USAF) was finally established in 1947, it was claimed that Billy Mitchell finally had been vindicated. The army and the navy had great traditions dating back to the Revolution, and some army units could trace their lineage to before the War of Independence. A problem for the new USAF was establishing an air tradition because so much of its history was bound to the army. The USAF "founding fathers"—Patrick, Foulois, Fetchet, and Mitchell—had come to the U.S. Air Service of the Great War and the post-war era from other branches of the army. Billy Mitchell began his career in the U.S. Army Signal Corps, where his record was superb. With Mitchell, however, there was a long history of being before the public eye—his war record, his feud with the navy, his constant flow of articles in popular magazines, and then the famous 1925 court-martial.

The end of World War II in 1945 did not guarantee peace, and as hostile feelings between the United States and the Soviet Union grew deeper it was imperative to bolster support in America for the new, expensive, and perplexing international role that the nation had to assume after the war. Under President Harry S. Truman, himself a veteran of World War I, the Defense Department was established and the U.S. Air Force was created. The missions assigned to the new service were vast and complex, and the Strategic Air Command became the symbol of the new power and the role of the United States. It was important to find those who had spoken out forcefully for the national defense, for preparedness, and for airpower before Japan's attack on Pearl Harbor. In the minds of many, especially those who wore the new Air Force blue, Billy Mitchell was such a man. In 1955, at the height of the Cold War, Republic Pictures released the movie *The Court Martial of Billy Mitchell,* starring the great actor Gary Cooper, a man who had won an Oscar in 1941 for his role in *Sergeant York.* The end of this fanciful and inaccurate film depicts a convicted Billy Mitchell, in civilian clothes, leaving Washington. In the sky he sees a flight of four De Haviland aircraft, and, as they fly by, they change into modern jet fighters.

Billy Mitchell as acted by Gary Cooper, however, bore little resemblance to the actual person. In the movie Mitchell continually resists, with righteous anger, any suggestion of involving the press and politics in his case. The reality was just the opposite; Mitchell was a master at using the press and creating a public persona. The movie led the audience to believe that only after the crash of the airship *Shenandoah* in 1925 did Mitchell in righteous indignation speak out. This was a glaring inaccuracy. During Allen Gullion's cross-examination of Mitchell, Gullion—played brilliantly by Rod Steiger—brings up a number of Mitchell's predictions about the future technological advances in aircraft, such as transoceanic flight, missiles, and supersonic aircraft, giving the impression that only Mitchell had this vision of the future in the air. According to the script, Mitchell alone saw the need for a unified air service, though in truth Mason Patrick, Benjamin Foulois, and James E. Fetchet had gone on record supporting such a concept. The impression given to movie audiences was of Mitchell as a solitary Moses, dedicated to leading the American people out of a wilderness created by the hide-bound, unimaginative, anti-airpower generals of the War Department.

The movie's depiction of Mitchell's court-martial also bore little relation to reality. Viewers were given the sense that the court-martial was basically a kangaroo court, hell-bent on the destruction of Mitchell because he dared to question the General Staff and the War Department. Mitchell's enemy, a fictitious General Gutherie, played by the fine character actor Charles Bickford, was seen as making all of the rulings when, in fact, Colonel Blanton Winship, a Judge Advocate General officer of impeccable reputation, ruled on matters of law—and often ruled against the prosecution. When the script was sent to the Pentagon in June 1955, a number of army lawyers tried to point out glaring errors of fact as far as procedure was concerned. By and large they were ignored.[1] Members of the Office of the Judge Advocate General were invited to attend the premier showing of the movie at the Anacostia Naval Photographic Center in Washington, D.C. They were no less taken aback at the factual errors in the movie.[2] But by then the movie was in theaters, and Billy Mitchell, or perhaps Gary Cooper, was confirmed as a "prophet without honor."

The independent air force was created, and with good reason and fine results. Mitchell had indeed been on the mark when he warned about a potential attack on the Hawaiian Islands and the Philippines. Like many officers who had service in the Pacific area, Mitchell knew of the growing competition and hostility between the United States and the empire of

Japan. Billy Mitchell had seen the introduction of the automobile, airplane, radio, and motion pictures, and by reading records from his Virginia home one gathers that his wife had what labor-saving devices were available in the 1920s and early 1930s. A man of intelligence and good education from a first-rate institution did not need much prompting to make predictions. Mitchell understood the principle of mass, and his assembling of over 1,400 aircraft for World War I's September 1918 St. Mihiel offensive achieved air superiority over the battlefield and contributed greatly to the success of the first great American offensive operation carried out by John J. Pershing's AEF. The movie-makers sadly left out this great accomplishment, an act of planning and execution that won personal praise from Black Jack Pershing himself. The movie includes a perfunctory mention of Mitchell's many decorations (during the beginning of Gullion's cross-examination of Mitchell), but the reasons for receiving such high honors from the United States were omitted, leaving the audience in the dark as to who Mitchell actually was and what he had done prior to the 1921 sinking of the *Ostfriesland* and the 1925 crash of the *Shenandoah*.

What the movie did was to obscure Billy Mitchell, the man. It gave him the mantle of great patriot and the cloak of martyrdom. Without some understanding of the human Billy Mitchell, the script writers, producers, and directors were left with defense counsel Frank Reid's conspiracy theory — that "they" were out to get Mitchell because he dared to challenge the star-spangled generals of the War Department. One of those dark forces behind the projected martyrdom of Billy Mitchell was President Calvin Coolidge, who is seen in the movie directing the prosecuting lawyer, Colonel Sherman Moreland, to "try harder" to convict Mitchell. Moreland was portrayed by Fred Clark, an actor who had won accolades as a comedian on the screen, and, in keeping with his Hollywood persona, was regarded as a buffoon, constantly challenged and defeated by Reid, played by Ralph Bellamy, well-known as a very serious actor.

Though he was a great and talented actor, Gary Cooper portrayed only a fraction of the personality that was Billy Mitchell. Cooper was bound by the script, which was filled with historical errors and misrepresentations of the facts. The movie was great propaganda at a time when the United States needed underpinnings for its new world role, but in the long run it fixed in the American mind a vision that Gary was Billy, and nothing could have been further from the truth.

In history it is difficult to separate truth from myth, simply because so many prefer to believe the myth. Billy Mitchell was one of those figures in history who loomed at times larger than life, and his reputation was enhanced by a fledgling United States Air Force and reinforced by a movie in 1955. There was a public Mitchell persona: the daring airman, the crusader for airpower and an independent air force, the persecuted officer who dared to speak out, the fine Virginia squire horseman and hunter. Beneath the surface was a man haunted by the memory of his Senator-father; the lonely lad at Racine College; the tentatively defiant, young, imperialist-ordinated officer; the man always in need of money to maintain his way of life.

No one could ever take away from Billy Mitchell his great achievements in World War I or his crusade for an independent air force. But one must also recognize that the founders of American airpower were many, and most of them worked quietly within the army system. The magnitude of Billy Mitchell's personality, his literary achievements, his overactive ego, and his self-promotion obscured those who shared his quest. The life story of Billy Mitchell spans the time from the Spanish-American War to the Great War, from William McKinley to Franklin D. Roosevelt. He was a product of his family and his times—a family that left an indelible mark upon him. He, in turn, left his impression upon the history of the United States in the twentieth century.

NOTES

1. Memorandum, Public Information Office, Department of the Army, 27 June 1955, and Comments, Military Justice Division, Opinions Branch, 27 June 1955, in National Archives, Archives II, College Park, MD, Records Group 153, Records of the Judge Advocate General, General Courts Martial, William Mitchell Case Records, Entry 40, Case Number 168771, Carton 9214-21.

2. Memorandum, Judge Advocate General, Opinions Branch, 1 November 1955, in National Archives, Archives II, College Park, MD, Records Group 153, Records of the Judge Advocate General, General Courts Martial, William Mitchell Case Records, Entry 40, Case Number 168771, Carton 9214-21.

ONE

YOUNG WILLIE

T HE WINTER OF 1935–1936 WAS A HARD ONE, EVEN FOR
Wisconsin. Snow had covered the ground and piled in drifts, making
travel hard for those required to move about. At the Milwaukee train station
the station master awaited the arrival of a train that had been slowed by late
February snowfalls. Waiting with the station master was a hearse driver and
a few Great War veterans from a local aviators' post of the American
Legion. The train pulled into the station, and the veterans came to attention.
The first man off the train was Eddie Rickenbacker of the old 94th Aero
Squadron, the Hat-in-the-Ring Squadron, the greatest ace of the Great War.
Following Rickenbacker came a drawn-faced brunette, an attractive
woman, who walked with the assurance of one who had mastered horses in
the ring and who had weathered the onslaught of newspaper reporters'
countless questions. Two other women who had come to the station
embraced the widow, and all waited for the oak coffin to be unloaded from
the train.

The wooden box contained the earthly remains of William "Billy"
Mitchell, grandson of a congressman, son of a senator, war hero, general,
court-martialed and resigned officer. It had been Billy Mitchell who organized 1,485 aircraft for John J. Pershing's first great American offensive of
World War I at a place known as St. Mihiel, Mitchell who had challenged
the U.S. Navy over airpower in 1921 and sent the German battleship
Ostfriesland to the bottom of the sea with a few minutes of aerial bombardment. His court-martial in 1925 had been justified in the eyes of some, but
for many Mitchell had emerged as a martyr for the air defenses of the
United States. For his final trip home Mitchell had been dressed in civilian
clothes. Had he worn a uniform, he would have displayed many ribbons,
among them the Croix de Guerre, the Distinguished Service Cross, and the

7

Legion of Honor. But there was no flag on his coffin, no honor guard for a dead general save for Captain Eddie Rickenbacker and the aging veterans of the Legion.

When Mitchell flew back to Langley Field in 1921 after the sinking of the *Ostfriesland* there had been a band, and his flyers carried him on their shoulders. Now there were no soldiers, no representatives of the army that he had served during two wars. When Mitchell left Wisconsin for the Spanish-American War in 1898, few noticed the nineteen-year-old blond lad boarding the train, and just as many took note of his return thirty-eight years later. But what a flight it had been!

* * *

It was a long way from a cold train station in Milwaukee, Wisconsin, to Nice, France, but that was where William "Billy" Mitchell was born on December 29, 1879. His father, John Lendrum Mitchell, and his mother, Harriet Becker Mitchell, were vacationing in Southern France to escape the harsh Northern European winter when William was born in an apartment owned by M. E. Corinaldi. The house, known as the Maison Corinaldi, was situated on Place Grimaldi, a pleasant square. Two days later the child was registered at the Civil Office in Nice, France, with Corinaldi and a well-known local baker, a Monsieur Diederich, as the required witnesses. The father, a studious man, signed the registry and returned to the apartment on the Place Grimaldi where his wife, having borne him a first child, was recovering. Nice had not been affected by the devastating Franco-Prussian War of 1870 or the 1871 draconian Treaty of Versailles. The ravages of the Paris Commune, which steeped the City of Lights in blood, did not alter the slow pace of life in Southern France, and that suited John Lendrum Mitchell quite well. Nice was a small city of beautiful vistas and cuisine that blended Italian, French, and the sea. It was a good place for John L. Mitchell to be because the ambience lent itself to reflection and perhaps a little indulgence, a good combination to nourish the intellect, and there was nothing that he prized more. John Lendrum, Harriet, and the young William would not return to the United States until 1881.

There had always been distance as far as the Mitchell men were concerned. John's father had been a hard-working Scottish immigrant. Born on October 18, 1817, Alexander Mitchell arrived in frontier Wisconsin in 1839, nine years before the territory achieved statehood. With only $200 to his name, Alexander Mitchell began the Marine Fire and Insurance Company in Milwaukee and built the company into a thriving, well-respected business. During the financial panic of 1857, Alexander Mitchell kept his company solvent, assuring investors that the industrious Scot was worthy of their trust and further investment. His second wife had blue blood, being president of the Colonial Dames and one of the women who saved George Washington's estate, Mount Vernon, from decay and ruin.

It was into a world of new wealth and old blood that John Lendrum Mitchell was born on October 19, 1842, in Milwaukee. From an early age, however, John Mitchell preferred the scholarly life, a life that only money could provide. Often this placed a strain on the relationship between father and son, with Alexander Mitchell not really approving of his son's intellectual bent nor his interest in such public affairs as education and secondary schools. Alexander Mitchell's business abilities made him a wealthy man, worth over $20 million by the end of the American Civil War. He had the wealth to go to New York City to buy art for a magnificent new home, and he bid against John D. Rockefeller for art and won. Alexander would use his money and power to end his son's disastrous first marriage and help him arrange a second one.

In 1858 John Mitchell left Milwaukee for Europe to study at various universities. His great loves were literature and languages, and he excelled in both. During the summer of 1859 he was in Paris, taking a break from his studies in Dresden, Germany. The French emperor had just involved his country in the Italo-Austrian War and had commanded his troops on the battlefield. When word reached Paris of the victory on the bloody battlefield of Solferino, the capital went wild with celebration. John Mitchell was in the crowds watching Empress Eugénie and the Prince Imperial lead a procession to Notre Dame Cathedral to give thanks for the triumph of French arms. A few months later, when in Paris again, John watched as Napoleon III, or "Napoleon the Little," as he called him, returned in glory to his capital. Mitchell had little use for military pomp, and he had less regard for the regime of Napoleon III.[1]

Returning to Milwaukee in late 1860, John Mitchell faced the most defining moment in his life. In 1861 the United States went to war with itself, and he decided, against his father's advice, to join the Union army. He enlisted in the 24th Wisconsin Volunteer Infantry, which was being formed in Milwaukee. There John Mitchell found an old family friend, the newly commissioned First Lieutenant Arthur MacArthur. MacArthur's father, known as Judge for his years on the bench, was an old friend of Alexander Mitchell. The two men had been drawn to each other due to their Scottish heritage and their views on commerce and business. Like Alexander Mitchell, Judge MacArthur tried to curb his son's martial ardor but failed, and in 1862 the 24th Wisconsin marched off to war.[2]

Neither Arthur MacArthur nor John Lendrum Mitchell could ever have envisioned the links that would develop between the two families. Arthur MacArthur's son, Douglas A. MacArthur, would go on to be a legend in American history and would sit in judgment on John L. Mitchell's son, Billy Mitchell, in the spectacular 1925 court-martial that ended the latter's military career. In 1926 Billy Mitchell would resign, with great fanfare, from active service. MacArthur would be known as the conqueror of Japan and the commander of American ground forces in Korea until he was relieved of command by President Harry Truman in 1951. Both Billy Mitchell and Douglas A. MacArthur ended their careers because they chose to speak out against what they perceived to be ill-conceived, incorrect government policies. The two sons were never close friends, but they knew each other socially in Milwaukee, and Douglas MacArthur dated Harriet, one of Mitchell's sisters, for a short time. Billy and Douglas's time together was spent mainly at parties and receptions for Billy Mitchell when he returned from Cuba on his way to an assignment in the Philippines.

While John L. Mitchell was serving in Union Blue, his father increased the family wealth by building the Chicago, Milwaukee, and St. Paul Railroad. With over five thousand miles of track in use, Alexander's cars hauled military goods, manufactured and agricultural products, and troops. Alexander also purchased a country estate known as Meadowmere and spent more time dabbling in Republican Party politics.[3] In 1864 John L. Mitchell, who was serving with the Ordnance Corps, left the army and returned to assist with the family business.[4] War had changed John's outlook on life in that he developed an abhorrence of violence, even in sport. He spent more time in public matters in Milwaukee, especially in the area of education for all citizens. To please his father and in keeping with his personality, John L. Mitchell went into the family banking business, which was booming because of the war, but he never lost his burning desire to continue his education in Europe.

Wisconsin and national politics interested father but not son. Alexander Mitchell had been involved with the Whig Party, but with the rise of the Republican Party he became a supporter of Abraham Lincoln in 1860. The new Republican Party promised support for railroads and internal improvements, positions the Midwestern businessman could well identify with. But after the death of Lincoln in 1865, Alexander questioned the Radical Republicans' obsession with reconstruction policies. He switched to the Democratic Party and openly defended President Andrew Johnson during his impeachment trial.

On October 20, 1865, John married Bianca Coggswell, of a prominent Milwaukee family with great political connections (her father was the law partner of U.S. Senator Matt Carpenter of Milwaukee). Within the space of a few years Milwaukee society became aware that the Mitchells were constantly arguing, even in public. In 1871 Alexander began construction of a magnificent mansion for the couple in hometown Milwaukee. John and Bianca lived there in splendid surroundings, but the marriage continued to deteriorate despite the birth of a son, David, in 1875. In July 1877 John Mitchell ordered his wife to leave the mansion without David. She did so, but then publicly accused John of "habitual intemperance," a charge that he in turn leveled against her. Both sides sued for divorce, giving rise to even more gossip in Milwaukee society, but Bianca was not through with the Mitchells.

Bianca Coggswell Mitchell decided to fight every inch of the way for a divorce and a settlement. The couple had two living children: ten-year-old Alexander, who was judged to be "feeble minded" and was under the care of a physician at what was essentially an asylum in Massachusetts, and two-year-old David, who was by all accounts a bright child. It was David who was living in the Mitchell home when the divorce proceedings began in early August 1877. During the trial Bianca claimed that she had been a faithful and dutiful wife, but John, she charged, remained in a "state of habitual drunkedness." During these drinking bouts John became abusive both verbally and physically and refused to allow her any money for clothes or other necessities, giving her only $2 at a time. Bianca then named several women, including David's nanny, who had had affairs with her husband. She also identified the places and the dates John had, in the words of the court's findings, "had carnal knowledge of [their] bodies."[5] Bianca stated that on July 7 her husband, in a drunken state, drove her, without David, from the family's Milwaukee mansion. The *Milwaukee Sentinel* called this "The Great Scandal" and published a series of nasty letters between John and Bianca's uncle that clouded the issue even more.[6] Bianca asked her late father's law partner, Matt Carpenter, to represent her as the case dragged on with more details of the couple's unhappy life reported in the newspapers.[7] Even editorial writers urged that for the "sake of public moral," the case be speedily taken out of the public view and settled quietly.[8]

Late in the summer of 1877, however, Bianca sneaked into the mansion and tried to spirit the child away. She made her way back to a hired carriage but fainted as Alexander Mitchell and his faithful servants subdued the distraught and struggling woman. Alexander Mitchell had had enough; he simply told John not to contest the divorce, which was filed on the grounds of "cruelty and inhuman treatment." The divorce was granted on

September 3, 1877. It was Alexander who finally got Bianca to agree to a yearly annuity of $2,000, but David would remain with the Mitchells under the care of Alexander's wife.[9]

John Mitchell remained distant in his relationship with David, and care and affection for the boy passed to John's mother, Martha Mitchell, who made the boy her companion for life. When she moved to the warmer climate of Florida after the death of her husband she took David with her. Bianca Coggswell Mitchell lived on in Milwaukee in a comfortable boarding house, but depression haunted her. She taught German, French, and music and maintained close contact with many members of Milwaukee society who remembered the lurid details of her divorce. On November 2, 1882, she became very ill with chest pains, and her doctor told her she was suffering from a fatal heart condition, with only days to live. In delirium, she at first refused to have her mother, who was living in Chicago, called to her bedside, but then agreed that her mother and her best friends should indeed be there. On November 3, 1882, she died, leaving behind a short manuscript for a children's book entitled *Rigdom Funnidos*.[10] There is no evidence that John Lendrum Mitchell acknowledged the death of his first wife, nor did he or David attend her funeral.

After the messy public divorce, John traveled in order to "regain his health," and when he finally returned to Milwaukee he seemed rested and ready to begin again. The family wealth increased as John became associated with various banks, and his involvement with various organizations added to his social prominence. He was an active member of the Grand Army of the Republic (GAR), a politically powerful organization that gave him a standing in the Wisconsin Democratic Party. In his large home in Milwaukee he constantly received old army comrades as well as members of the business community. In 1871 Alexander Mitchell was elected to Congress as a Democrat and served two terms.[11] The Wisconsin Democratic Party chose Alexander as their nominee for governor, but he declined to run, preferring to remain as president of his railroad. By 1878 John had married Harriet Danforth Becker, a member of a socially prominent Cooperstown, New York, family, and he decided to return to Europe to study and to visit historic sites.[12]

Life in France in 1878 and 1879 was unsettled. France had just been soundly defeated in the Franco-Prussian War of 1870 and had suffered severe domestic violence during the Paris Commune of 1871. Politics remained chaotic as the French established the form and structure of the new Third Republic. Critical elections in October 1877 gave the Republicans the necessary mandate to continue to build their government, but conditions were still unsettled by the time the Mitchells arrived in

France. With Harriet pregnant with their first child, it was natural that they would seek the less politically volatile climate of Nice in Southern France. It was there that William Mitchell was born and registered. The Mitchells remained in France for almost three years, and William, now nicknamed Willie, learned French, using it with the same facility that he did English. His father insisted that his son learn to speak the language, since he himself was an accomplished linguist.

There was still tension in the air when John and Harriet Mitchell returned to Milwaukee. John's father was not pleased with his three-year sojourn in Europe, and there was no indication that he was particularly affectionate toward his new grandson.[13] From 1884 to 1885 John Mitchell served as head of the Milwaukee Public Schools, and he continued to host members of the GAR and the local and state officials of the Democratic Party, which was becoming more reformist. The home in Milwaukee and the country estate at Meadowmere became centers of art and culture, reflecting the stimulating intellectual climate in Europe in the early 1880s. The Mitchell family grew with the addition of another son and three daughters. It appears that the children were not well disciplined, and William Mitchell had a way of being intrusive and demanding. A Scottish nanny, Mary Alexander, was employed to attend the children, but Harriet Mitchell emerged as the strongest influence in the lives of her children.[14]

In 1884 the Milwaukee County and City Democratic Committee was locked in a squabble over leadership, and John Mitchell was selected chairman of the committee for upcoming elections because he was believed to be the only man who could unite Democrats behind a reformist platform. Mitchell's work with the Milwaukee public schools and public libraries made him a man acceptable to all factions.[15] He was well thought of by the GAR, and although its members normally voted for the Republican Party, in this case they would vote for one of their own regardless of party. In February he had taken William to Martha Mitchell's "orange plantation," Villa Alexandria, in Florida. He had decided to remain for several months, since young William enjoyed the outdoor life.[16] There were no hints of political activities in his letters either to his father or to his wife. There is no evidence that John spent much time with his son David, who also lived at the villa, or that William was encouraged to be close to his half-brother.

After the death of Alexander Mitchell in New York City on April 19, 1887, John was free to devote himself to matters beyond banking and business. Now politics attracted him, and he was elected to Congress as a Democrat in 1891, two years later becoming a U.S. Senator. He began his senatorship as Grover Cleveland, a Democrat with reformist tendencies, was sworn in as the nation's twenty-fourth president. John rose quickly in the party, becoming a national committeeman, and he attracted the attention of William Jennings Bryan, himself a rising star in the Democratic Party. Mitchell purchased a modest home on Capitol Hill but sent William to a boarding school rather than bring him to Washington. At age nine, William Mitchell entered Racine College, an Episcopal school in Racine, Wisconsin.

When Willie Mitchell arrived in Racine, the college consisted of six buildings: a chapel, dorms, residences, and school buildings. Founded in 1852 by the Episcopal Church, Racine College aspired to be the "University of the Northwest," but by 1890 it had made little movement beyond being a preparatory school for Episcopal boys.[17] It offered, however, a full range of classes, including religious instruction, and in 1893 Mitchell was confirmed into the Episcopal Church as a matter of course. Living at Racine College would be a period of difficult transition for William because he had received scant discipline at home and was used to riding at Meadowmere and hunting in various areas of Wisconsin. By age twelve he had shot and had stuffed almost two hundred different types of birds.[18] To complicate matters, his mother and father were living in Washington, D.C., and visits with his family would be limited to Christmas and summer vacations. The young William Mitchell would prove difficult to handle at the sedate Racine College.

The young William was an average student who spent more time at sports than the faculty would have liked. The picture that emerges over the years is of a young boy who had breaches of discipline that were not serious, a fair student who did well in history and languages, but also a boy who harbored a growing resentment at being at Racine College while his parents made their home in Washington. Any visit by his mother was an event for William, and he continually asked Harriet to return to the school to see him.[19] Willie depended on his mother as a correspondent and a source of support while he attended school. Indeed, most of the letters from teachers and administrators informing the Mitchells of their son's academic, social, and athletic progress were addressed to Harriet Mitchell.

When Harriet received a report card stating that William's conduct was "perfect" she wrote back to inquire as to what that meant, as his previous report card had contained only a "fair" mark. Watson R. Hake, a college official, responded, writing that the earlier mark was for such minor offenses as talking during the saying of grace before the meal and loud talking during the meal, certainly nothing serious. Hake added that Mitchell was attentive to his studies and had such a personality as to be a favorite of the academic instructors.[20] William was forming a distinct personality, with an ability to be charming but very assertive, and with a tendency to resist authority. As a case in point, the school had a policy of physical training that required the boys to run. William implored his mother to write to the headmaster to excuse him from the exercise, claiming that they ran five and a half miles with no rest, and he was afraid that such running would make him sick.[21] Another part of William Mitchell's personality—the ability to exaggerate situations, using it to his advantage—emerged and would be a part of Billy Mitchell for life.

John Lendrum Mitchell's political career was on the rise, but his relationship with William mirrored his own distant relationship with his father. He visited William at school one day in April 1891—from 9 to 11 A.M.—but left for Washington prior to lunch. At the time, William was suffering with conjunctivitis and planned to see a doctor about the malady. In a curious letter to his mother, William told her that his father would write to her from Chicago, where he had political business, and discuss their son's eye problems. It seemed that William expected to make his own appointment with an eye doctor, probably with the assistance of the authorities at Racine College, but there seemed to be no sense of urgency or concern.[22]

During William's second year at Racine College, his main teacher Arthur Piper wrote to Harriet Mitchell that he believed young lads who had a life of affluence needed more "repression and steady government" than other boys. Piper was convinced that William Mitchell would be withdrawn from Racine College, and he wanted to advise William's mother about his future.[23] This had to come as a surprise to Harriet, as there had been no serious discussion of Willie's leaving Racine College. Mitchell's parents invited Piper to visit them in Washington over the Christmas break, and the teacher later wrote that young Willie had spent a great deal of time regaling them with stories about duck hunting along the Potomac River and in Virginia.[24]

William Mitchell returned to Racine College for the fall term of 1893. Now entering his teenage years, Mitchell threw himself into sports, often to the detriment of his academic studies. He informed his mother that during fall sporting events he had received medals for his performance in the broad jump, pole vault, ball throwing, and the 100-yard dash, and he broke a school record in the broad jump.[25] By the end of October 1893, Willie was ready to compete in bicycle races, and he appeared to be enthusiastic over the upcoming school athletic competition. But while writing his mother about his athletic prowess he also asked if she could visit him at the college prior to Christmas. He ended a letter stating that, "I suppose the house is all in order now and the children will be there soon."[26] For the young

William Mitchell it appeared that, regardless of his father's exalted political position, he would not be going to the home on K Street in the nation's capital.

The year 1893 was not a good one for Senator Mitchell. The financial panic of 1893 had forced him to briefly close the bank in Milwaukee, but he was determined that no one should lose money in a Mitchell institution. He pledged to support the bank with almost $1 million of his personal wealth. It is highly probable that by supporting the bank in 1893, John L. Mitchell decreased the family's wealth considerably.[27]

The financial crisis of 1893 was more severe for John Mitchell than anyone really knew. Because Mitchell was serving his term in Washington he could not deal with the day-to-day management of one of the family's most important sources of wealth. Washington Becker, Harriet's brother and a Milwaukee businessman, had to step in to stop the severe drain on the bank and on his brother-in-law's finances. In July 1893 he recommended that the entire structure of the bank be reorganized to end unsound banking and loan practices.[28] By September John Lendrum Mitchell realized that he was in deep trouble: a grand jury convened to look into the operations of the Mitchell bank. To forestall further problems, John authorized Harriet to sell some of the best horses in the stables at Meadowmere.[29] At one point Harriet wrote to John, "We will probably lose all our property anyway and may as well yield gracefully."[30] In fact, John Lendrum decided to avoid returning to Milwaukee because, "I would have no end of papers served on me and there is no telling what the grand jury might do."[31] The bank battle went on, with Mitchell doing everything possible—even renting the Mitchell mansion to the Milwaukee Deutscher Club and selling more prize horses from Meadowmere—to save the Mitchell bank and his own reputation. Under Becker's guidance, the bank survived the crisis, and the *Milwaukee Sentinel* announced in July 1894 that the bank had balanced its books and was open for business. But the 1893 banking crisis had seriously drained Mitchell's finances and forced Harriet Becker Mitchell to use her own personal fortune to keep the family going.

To complicate matters for Harriet and John Mitchell, Martha Mitchell was becoming mentally unstable. In 1888 she permanently moved with her grandson David to Villa Alexandria, near Jacksonville, Florida, with the understanding that she would receive a yearly stipend from her late husband's estate. In November 1888 John Lendrum received a letter from his uncle Harrison Reed, once an editor of the *Milwaukee Sentinel* and now a Florida businessman, that Martha was now "of unsound mind."[32] Throughout 1889 Reed and David Mitchell kept John informed about his mother's condition, which deteriorated to a point where she threatened to sue her son over her stipend.[33] That was avoided, but the situation festered until September 1893, when again she threatened a lawsuit over John's failure to send her money. The possibility of more court action, in the middle of the financial panic that affected the family's Milwaukee bank, could not have come at a worse time. This time she demanded a lump sum payment of $10,000, for which she would give up any claim to a "future annuity."[34] Martha Mitchell received her sum, and John and Harriet Mitchell essentially severed ties with the residents of Villa Alexandria. Will, who then was at Racine College, had some idea of the family difficulties with his grandmother, but his age makes it difficult, if not impossible, to gauge the effect of this family crisis except by his lack of interest in visiting the Florida orange groves. Martha Mitchell died in Jacksonville, Florida, on February 14, 1902.

In November 1893, William wrote another letter to his mother telling her of the athletic competitions he had won, carefully drawing the medals he would receive.[35] Medals and recognition were obviously important to him, as they likely would be to any fourteen-year-old boy, but they also compensated for his sense of isolation from his parents and siblings, now installed in the home on K Street in Washington. He also wrote to his father about his athletic interests, and his father responded in a stilted letter: "I quite agree with you about the game of cricket. It seems to be as exciting as baseball." The senior Mitchell then wrote, reflecting his feelings about violence, "Football is the rage just now. A raging game it certainly is—too much so for bodily safety. It is a grand thing to cultivate muscle, nerve, courage. But football, as played at present, is too savage." He also stated, "Your letter is silent as to your studies. This comes of over-modesty, I suppose."[36] With no mention of the prized medals, John Lendrum Mitchell's letter to his son was basically an admonition, dismissing his son's letter with only a comment, and a question about studies that was sarcastic.

If there was a male in William Mitchell's life it was his "Uncle Doc," Volney D. Becker, who lived in Milwaukee. He would make the short trip south to Racine periodically to visit Willie, and while there they would talk about hunting and fishing. Uncle Doc also would listen to his nephew's complaints about being at Racine College while the rest of his family resided in the nation's capital. It was Uncle Doc who seemed to understand that Willie was becoming an active teenager. Willie would mention each of his uncle's visits in his letters to his mother and appeared to enjoy their time together.[37]

William Mitchell now felt, and with good reason, that he was not receiving the financial support he thought he should from his family. William was having difficulty in physics and needed a tutor, and he wrote his mother very pointedly that it would not cost the family anything to get one at Racine College. At the same time he began a campaign to transfer to Columbian University in Washington, D.C. He claimed that he was preparing by mastering Caesar, Cicero, and Virgil as well as algebra, German, history, and other subjects he would need for admittance into Columbian. He ended one letter by asking his mother to send him a box of food for Thanksgiving because what little money he had did not allow for many treats. One particular request was for a baked possum![38] It seems William's claims of course mastery were not true, for H. D. Robinson of Racine College wrote Harriet Mitchell that "I sometimes wish that he would put more time on his lessons for I know that he can do better work. Yet I would not have him take less interest in his out-door sports."[39]

After the Christmas holidays and the arrival of report cards, reality set in, and Mitchell tried to explain that illness had caused absences that, in turn, brought about low grades. Willie also asserted that his teachers were not very good, but he was sure that if he attended Columbian he would do much better.[40] That spring his letters returned to a familiar theme—the need for an allowance of some sort. "Will you please send me some money," he wrote his mother, "[because] all the fellows get money from home. . . . Be sure to send me some money."[41] A few days later Willie wrote to his mother, "I am sure that papa got it when he was away at school. There are lots of fellows here who are a lot poorer than we are and they get it. I don't have any money at all and everybody else has money, even little kids about 11 or 12."[42] Spending money and a desire to attend Columbian in Washington occupied William Mitchell's letters to his parents, who were, by necessity, indeed parsimonious with their son. These sources of frustration would have a definite impact on William Mitchell's future. Much depended, of course, on the precarious state of the Mitchell family finances after the panic of 1893.

John Lendrum Mitchell's political career was going well. He had been selected as chairman of the Senate's Banking Committee, and populists and reformers urged that he be considered as a possible vice-presidential candidate in 1896, a good complement to the rising power in the Democratic Party, William Jennings Bryan. During the summer months the Mitchells often took their family—William included—to Europe. While there, John Mitchell manifested another side to his complex personality: his hatred of war, born during his Civil War experiences, was balanced by a love of military history. The entire family toured many of the Napoleonic battlefields, much to the delight of Willie and much to the disgust of his sisters. Indeed, William Mitchell's interest in military history can be attributed to his father's appreciation of the subject and his early exposure to the battlefields of Europe.

In the spring of 1894, before one of his family's European vacations, William made a good case for admission to Columbian University's School of English. William had obtained a college catalog and carefully outlined his preparation for admission to a university with very high standards. He would have to enter Columbian's preparatory school for a year, but he assured his father that he more than met the requirements in Latin, English, history, and arithmetic.[43] Though William's parents had at one time discussed Harvard University as the potential institution for the continuation of their oldest son's education, Columbian was admittedly a more realistic choice. William Mitchell convinced his father, and it was agreed that in 1896 he would attend school in Washington, D.C.

William was not one of Columbian's stellar students. It was obvious that he was highly intelligent and had the benefit of a good preparation at Racine College, but academics had taken second place to athletics. John Mitchell received his son's grade report in late May 1896 and found that in Latin his son had earned an A, but in logic, chemistry, and philosophy he was only able to garner Cs. William Mitchell did spend a good deal of time in sports, and he was becoming a superb horseman. Like many a young man raised on tales of the Civil War, Mitchell was well aware of the growing conflict with Spain, and this probably detracted from his studies.[44]

In 1895 the War for Cuban Independence flared up after a period of twenty years of peace in the Spanish colony. During the election of 1896 Republican Party candidate William McKinley indicated an interest in the situation in Cuba. This issue was very well publicized in the newspapers owned by William Randolph Hearst, who favored aggressive intervention in the Cuban struggle. Senator Mitchell, like many Democratic Party reformers and intellectuals, opposed any sort of American intervention, believing it would lead to certain war with Spain. However, on the night of February 15, 1898, an explosion sank the American battleship USS *Maine* in Havana Harbor. Two officers and 258 American sailors died in the mysterious explosion, and Hearst's newspapers were quick to lay blame for the destruction of the *Maine* on Spanish authorities. On April 25, 1898, Congress declared that as of April 21 a state of war existed between Spain and the United States.

Until that spring William Mitchell had evidenced no interest in a military career. Subsequent letters and statements by Mitchell indicate that he had considered a career in the family business or perhaps a post in the diplomatic service of the United States. But the lure of adventure was strong for a nineteen-year-old. He had seen the parade of GAR veterans in his father's home, had been raised with stories of the old 24th Wisconsin, and had been introduced to now Brigadier General of Volunteers Arthur MacArthur, the often-wounded recipient of the Medal of Honor, hero of the Civil War. Much to John Lendrum Mitchell's distress, his son announced that he would return to Wisconsin to sign up in his father's old regiment, now designated the 1st Wisconsin Volunteer Infantry. Very quickly the regiment was formed and ordered to report for training at Chickamagua, Georgia. At the same time Senator John L. Mitchell began the process of obtaining an officer's commission for his oldest son in the Wisconsin National Guard. A commission for one so young, even in the militia, would require political clout, but the father who hated war and the senator who spoke out against the conflict with Spain used his position to assist his son. William Mitchell became a second lieutenant in the 1st Wisconsin, assigned to the regimental signal company, and prepared to move to the Chickamagua training area.

On the day they left the train station in Milwaukee, "There were some pretty touching farewells paid when the boys left, with women fainting and crying all around like the world was coming down," he wrote his mother.[45] Willie's outdoor activities and his participation in sports had toughened him physically, and he found that he could deal with the rigors of military life with no difficulty. Mitchell was introduced to hardtack, corned beef, and coffee, a diet his Civil War veteran father would have very quickly recognized from his own war experiences.[46] He saw many of the 1st Wisconsin become ill, and he reported that a good many had died from exposure to the elements, poor camp sanitation, and from ill-preserved and ill-prepared rations.

Training at Chickamagua left much to be desired. The nation was simply unprepared for war of any sort. Uniforms, equipment, rations, transportation, ammunition, and all sorts of necessary equipment were either in short supply or simply nonexistent. In July the 1st Wisconsin moved near Jacksonville, Florida, where disease took a terrible toll on the troops. (Notably, though he was stationed near his grandmother's villa, there is no evidence that he ever visited.) The sickness rate was alarming, and Mitchell reported to his mother that in Company C of the 1st Wisconsin, which had 106 men on the rolls, only twenty-six soldiers were present for duty. The July heat was staggering for the Wisconsin regiment, unaccustomed to Florida's climate, but rumors that they were about to leave for duty in Havana, Cuba, kept morale fairly high.[47]

Mitchell chafed at not being in action, but he understood that he was rising in the estimation of his military superiors. Because of physical conditioning prior to enlistment, Mitchell was able to survive the unsanitary conditions in the camps around Jacksonville. As officers became ill due to bad water, spoiled rations, and heat, Mitchell was able to take up their duties because he was never ill.[48] Jacksonville had been designated as the site for the concentration of the VII Corps, which initially had no specific mission. However, as the war progressed VII Corps was designated to occupy Cuba. There was a great question as to whether to deploy the volunteer regiments to the island after hostilities. Eager for action, Mitchell was disgusted when it seemed that the regiments, now severely depleted because of typhoid fever and other diseases, would simply be sent home. For a youth lusting for true adventure this was frustrating for William Mitchell. He also was anxious that his father visit him in camp near Jacksonsville, hoping that it would be a matter of pride for his father, the old veteran of a fine fighting regiment in the Civil War, to see his son in uniform.[49] Mitchell was well aware of the positive comments his work had elicited from his commanders, many of whom were Regular Army veterans. With some aggravation, Mitchell wrote his mother that he did not think a visit to the camp would hurt his father at all. However, the elder Mitchell would not commit to a visit and that autumn had a severe bout of rheumatism that made travel impossible and raised doubts about his future in Washington. By August rumors were rife that the 1st Wisconsin would indeed see service in Cuba, and Mitchell informed his mother that he was more than ready to take his new sword, which he had recently purchased, to the island.[50] His mother also had sent him money to purchase a horse that would befit an officer's status.[51]

By the end of August, though, Mitchell believed that the Wisconsin regiment would go neither to Cuba nor Puerto Rico, and would miss the war. He had decided to affiliate with the U.S. Army Signal Corps, a move that would eventually bring him into contact with the air and the Air Service.[52] His parents had been upset that he had not finished his course of study at Columbian, and William wrote to assure them that he was learning more in the army than he ever would at the university.[53] By September, the army had decided to muster a large portion of the 1st Wisconsin, especially the sick, out of active service, but as far as Mitchell was concerned it did not matter because he was, for all purposes, in the Signal Corps.[54] Because of his age an official transfer and Regular Army commission were two years away. He proudly informed his mother that his captain had put him in charge of a section, important because it cemented his relationship with the Regular Army.[55]

It did not take Mitchell very long as a newly minted second lieutenant to criticize the War Department. "The war department are the people to blame right through and any body who has a grain of common sence [sic] can see it right straight through," he wrote to his mother. "And I do believe that if we had been up against a first rate power they would have whailed the mischief right out of us."[56] In the same letter William stated that he was considering leaving the service, even though he liked it. A large portion of the Wisconsin regiment had left for Milwaukee, and William was having feelings of regret seeing them depart for home.[57]

As September drew to a close Mitchell could look forward to a major change in his off-again, on-again military career. He had been given command of a signal company, and word reached the signal unit that it would indeed be going to Cuba. He asked his father to intercede in Washington to see that he was either sent to Cuba with the signal unit or transferred into the cavalry. As he wrote his father, he did not want to go back to Milwaukee without ever leaving the United States, saying "it would look kind of funny" to return home under such circumstances.[58] Senator Mitchell, who had decided not to seek a second term in the Senate and who was in opposition to President McKinley and the Spanish-American War, did not respond to his son, but it really did not matter. The War Department had decided to send the signal unit to Cuba, and it arrived on the island in late November. In fact, it was fortunate for William Mitchell that he did not transfer into the cavalry. He had expressed interest in joining a volunteer cavalry regiment that had been modeled on the famed 1st Volunteer Cavalry Regiment, better known as the Rough Riders.[59] The regiment Mitchell was interested in did not go to Cuba, and had he joined it he would have been demobilized with the regiment and sent home to Wisconsin, having seen no action. His unit occupied Camp Columbia, on the coast about seven miles from the city of Havana.

Nineteen-year-old Second Lieutenant Mitchell was euphoric. He was bivouacked near a camp of Cuban *insurectos,* and the colonel commanding the group, which was armed with a strange variety of old and new weapons, promised to give Mitchell a machete that, the Cuban claimed, had been used against the Spanish. The pomp and circumstance of the Regular Army intrigued Mitchell, and he told his father that he was thrilled at retreat when the regimental band played the *Star-Spangled Banner.* Perhaps remembering the record of the old 24th Wisconsin and that his father had "seen the elephant" during the Civil War, Mitchell told his father that Cubans living in the heavily forested area periodically shot at the Americans. " I think," he wrote, "that we will have to turn in and lick the Cubans before long."[60]

With Christmas approaching, Mitchell appeared to be perfectly at home in the field with the Signal Corps. He would shoot some wild birds for a Christmas meal, and the countryside yielded pineapples, figs, mangoes, and all manner of vegetable-garden items. To make matters even better, he still commanded 2nd Company, and his commander told him that soon he would undertake independent signal missions in the interior. Subsequent letters expressed no more ideas about joining the cavalry. His lengthy correspondence made no reference to military glory, but to action, exciting missions that promised to take place early in 1899. Also, it appears that young Mitchell sought the approval of his superior officers, and he continued to write his mother about every bit of praise. Mitchell sought out difficult missions that would be noted by his superiors. While he basked in the light of praise, he was still uncertain about his future plans. Periodically he mentioned returning to civilian life to finish his education and then go into the family businesses. He relied on financial support from his mother, but that was due to the low pay of a second lieutenant in the army. He would, however, continue to receive financial assistance for decades to come.

By January 1899 Mitchell had been in the army only nine months, but he had compiled a good record. His father, preparing to return to study in Europe, believed that his first son would return to Wisconsin to work in the family business. William Mitchell's newfound liking for the military and his determination to seek areas of action ran counter to his father's wishes, and though for several years to come William Mitchell would waiver between the calls of civilian life and the excitement of military service overseas, as he continued to serve and win accolades the lure of civilian pursuits would fade into the background.

NOTES

1. John L. Mitchell to Bob Chivas, from Dresden, Germany, 17 March 1860, in the John L. Mitchell and Mitchell Family Papers, Golda Meir Memorial Library Archives, University of Wisconsin, Milwaukee, Folder 16. (Hereafter cited as the Mitchell Family Papers.)

2. D. Clayton James, *The Years of MacArthur, Volume 1, 1880–1941* (Boston: Houghton Mifflin, 1970), 12–13, 64–65.

3. Ruth Mitchell, *My Brother Bill: The Life of General "Billy" Mitchell* (New York: Harcourt, Brace and Co., 1953), 26. This book must be used with great care in that Ruth Mitchell was devoted to her brother. The book does give the "Mitchell Family Line" in regard to Mitchell's controversial career.

4. U.S. Congress of the United States, *Biographical Directory of the United States Congress* (Washington: Government Printing Office, 1989), 1514.

5. *Milwaukee Sentinel,* 20 August 1877, 1.

6. Ibid., 21 August 1877, 1.

7. Ibid., 30 August 1877, 2.

8. Ibid., 3 September 1877, 2.

9. Article from the *Milwaukee Sentinel,* ca. September 1878, Mitchell Family Papers, Folder 38.

10. *Milwaukee Sentinel,* 4 November 1882, 10.

11. U.S. Congress, *Biographical Directory,* 1989.

12. Burke Davis, *The Billy Mitchell Affair* (New York: Random House, 1967), 12.

13. Ibid., 12.

14. Ibid., 12–13.

15. John Mitchell to Wife, 2 August 1884, Mitchell Family Papers, Folder 17.

16. Ibid., 14 February 1884, Folder 18.

17. Milwaukee *Daily Journal,* 1 December 1891, in the William Mitchell Papers, Library of Congress, Washington, DC (Hereafter cited as Mitchell Papers), Carton 19A.

18. Ruth Mitchell, *Brother Bill,* 25.

19. Mitchell to Mother, 3 April 1890, and Mitchell to Mother, 9 April 1890, Mitchell Papers, Carton 19A.

20. Hake to Mrs. Mitchell, 23 December 1890, Mitchell Papers, Carton 19A.

21. Mitchell to Mother, 26 February 1891, Mitchell Papers, Carton 19A.

22. Ibid., 21 April 1891, Mitchell Papers, Carton 19A.

23. Arthur Piper to Mrs. Mitchell, 7 November 1891, Mitchell Papers, Carton 19A.

24. Robinson to Mrs. Mitchell, 22 October 1892, Mitchell Papers, Carton 19A.

25. Mitchell to Mother, 2 October 1893, Mitchell Papers, Carton 19A.

26. Ibid., 30 October 1893, Mitchell Papers, Carton 19A.

27. Alfred F. Hurley, *Billy Mitchell: Crusader for Air Power* (Bloomington: Indiana University Press, 1975), 2.

28. Washington Becker to Harriet, 27 July 1893, Mitchell Family Papers, Folder 10.

29. John Mitchell to Wife, 27 September 1893, Mitchell Family Papers, Folder 20.

30. Ibid., 23 September 1893, Mitchell Family Papers, Folder 20.

31. Ibid., 2 October 1893, Mitchell Family Papers, Folder 20.

32. Reed to John Mitchell, 13 November 1888, in the State of Wisconsin, Federal Writers Program, Carton 27, Alexander Mitchell Folder. Provided to the Golda Meir Memorial Library Archives, University of Wisconsin, Milwaukee, by the State Historical Society of Wisconsin, Madison. (Hereafter cited as Alexander Mitchell Folder.)

33. A series of letters including one from David Mitchell to John L. Mitchell, 7 April, 17 May, and 22 May 1889. The letter concerning the lawsuit is from Reed to Mitchell, 24 May 1888, Alexander Mitchell Folder.

34. Reed to Mitchell, 29 September, Alexander Mitchell Folder.

35. Mitchell to Mother, 4 November 1893, Mitchell Papers, Carton 19A.

36. John L. Mitchell to William, 18 November 1893, Mitchell Papers, Carton 19A.

37. Ruth Mitchell, *Brother Bill,* 36.

38. Mitchell to Mother, 17 November 1893, Mitchell Papers, Carton 19A.

39. Robinson to Mrs. Mitchell, 20 December 1893, Mitchell Papers, Carton 19A.

40. Mitchell to Mother, 20 February 1894, Mitchell Papers, Carton 19A.

41. Ibid., 18 April 1894, Mitchell Papers, Carton 19A.

42. Ibid., 27 April 1894, Mitchell Papers, Carton 19A.

43. Mitchell to Father, 23 May 1894, Mitchell Papers, Carton 19A.

44. Grade Report Card from Columbian, 27 May 1898, Mitchell Family Papers, Folder 38.

45. Mitchell to Mother, 21 May 1898, Mitchell Family Papers, Folder 38.

46. Ibid.

47. Mitchell to Mother, 22 July 1898, Mitchell Family Papers, Folder 38.

48. Ibid., 14 August 1898 and 26 August 1898, Mitchell Family Papers, Folder 38

50. Ibid., 29 August 1898, Mitchell Family Papers, Folder 38.

51. Ibid., 3 August 1898, Mitchell Family Papers, Folder 38.

52. Ibid., 9 August 1898, Mitchell Family Papers, Folder 38.

53. Mitchell to "Uncle Doc" Becker, 3 September 1898, Mitchell Family Papers, Folder 38.

54. Mitchell to Mother, 3 September 1898, Mitchell Family Papers, Folder 38.

55. Ibid., 6 September 1898, Mitchell Family Papers, Folder 38.

56. Ibid.

57. Mitchell to Mother, 11 September 1898, Mitchell Family Papers, Folder 38.

58. Mitchell to Father, 20 September 1898, Mitchell Family Papers, Folder 38.

59. Mitchell to Mother, 13 October 1898, Mitchell Family Papers, Folder 38.

60. Mitchell to Father, 24 December 1898, Mitchell Family Papers, Folder 38.

TWO

FROM THE PHILIPPINES TO ALASKA

WILLIAM MITCHELL HAD EVERY REASON TO BE PLEASED with his prospects for 1899. What was left of the Wisconsin militia regiment was in Cuba, and signal units were told that there were a number of critical missions to be undertaken early in the year. Miles of wire would have to be stretched between military camps and the capital city of Havana, and Lieutenant Mitchell would be involved in exciting and probably dangerous work in the forests and jungles, where many Cubans had not accepted the American occupation of their island. "All of Cuba is ours and we who are here feel very much gratified at the ending things have taken. Everything went smoothly and little blood was shed. I think that only ten or twelve people were killed yesterday," Willie wrote his mother in January 1899.[1] He was referring to the final surrender of Spanish troops, which Mitchell witnessed at the Spanish governor-general's palace in Havana.

Unlike his father, William accepted the annexation of Spanish territory with enthusiasm. "[The Spanish-American War] marks the beginning of a new policy on the part of the U.S., that of territorial expansion and showing himself to the world as one of the greatest of nations," he wrote to Harriet Mitchell.[2] Certainly John Lendrum Mitchell did not share his son's pride in the newfound American empire, and he cautioned his son against a military career. Regardless of their differences, William continued to ask that his parents visit him in Cuba. His letters to his father judiciously avoided the jingoism he evidenced in letters to his mother—Mitchell wanted to avoid an acrimonious exchange of letters with his senator-father, who had made his position against expansion clear on the floor of the Senate when in June 1898 he stated:

And now the Philippines be upon us. The nation, shorn of its judgement, is led captive by its emotions. We are to establish ourselves permanently in the far East, and must have a coaling station in mid-Pacific as a basis for aggressive action. Under a passing stress of war we are to be pressed into taking a first step in imperialism—a policy which may benefit the favored few, but to the ordinary mortal it means the path to the barracks or possibly the poorhouse. All this at the precise time when we should avoid compromising ventures. Europe already questions our sincerity in the declaration touching Cuba. The seizure of Hawaii would remove any doubt as to our all-round land-grabbing intentions.[3]

Growing concerned over his son's soldiering, John Mitchell continued to press William about a career. William responded in one letter, "As to my going into the army, of course as I look at it now I would not make it my life's work. Although after I am in for a year or so it might look different. I would like a [Regular Army] commission as it stands now anyway."[4] He was quick to inform his father that the colonel of the VII Corps' Signal Battalion had singled him out for his work and indicated that he would be a prime candidate for further training at the newly established Signal School at Fort Myer, Virginia.[5]

An end to hostilities and the occupation of Cuba was a learning time for Mitchell. He thrilled at being given assignments to lay wire and establish lines between posts because he had an independent command of twenty-two soldiers, and there was always the possibility of action. In mid-January, however, his battalion commander assigned him and his command to clean up and reopen the Havana cable office. When Mitchell and his signalers arrived at the office they found, "It is hardly fit for a human being to go into. The sinks [toilets] apparently have the accumulation of ages in them and I got to work immediately."[6] Mitchell found out that not all military duties were exciting. Cleaning filthy toilets and preparing a disabused building were all part of the occupation experience. He hinted to his father that there might be more action in the future when he wrote, "Don't you suppose that if they have any trouble in the Philippines you can have me sent out there?"[7] Regardless of John Mitchell's anti-imperialist position, he would help his misguided son as much as he could. The now ex-senator was preparing for his return to Europe. He had planned to attend the University of Grenoble in France and would, in fact, remain there until 1902.[8]

Despite William's desire to go to the Philippines, he had been given an important assignment in early February: to construct telegraph lines between the capital of Havana and the city of Santiago. Wires laid by the Spanish had been cut or totally destroyed, and Mitchell, who now found himself in command of another lieutenant and a group of U.S. soldiers, had to organize the force, including Cuban laborers, to complete the mission as quickly as possible.[9] With an energy characteristic of the adult William Mitchell, the task was organized and work begun in eight days. Mitchell had to employ Cuban labor and procure rations for them at the cost of $1.60 per day, which was quite high for Cuba at that time. He requested and received authority from Colonel H. C. Dunwoody, chief signal officer in Cuba, to pay the costs.[10] With authority granted, Lieutenant Mitchell and his troops went to work with a subordinate lieutenant, an army surgeon, about forty soldiers, sixty-three pack mules and wagons, and a large number of Cuban laborers—a handsome command and considerable responsibility for a militia lieutenant who was not old enough to qualify for a Regular Army commission.[11]

Mitchell's attitudes toward the Cubans were fairly typical of upper class, white, Protestant Americans at the end of the nineteenth century. He was surprised that "ignorant people" hid their smallpox-infected villagers from the army surgeon. "The Spaniards," Mitchell wrote, "are so far superior to them [the Cubans] as the Romans were to the Gauls. The only way to settle this thing is to annex the whole outfit [and] fight them until they dare fight no more."[12] Mitchell had traveled in Europe with his parents and was firmly wedded to European culture and mores. His homes in Milwaukee and Meadowmere were filled with European art and furnishings. He became, and remained, firmly convinced of the superiority of the Western world. By the second decade of his life William Mitchell was a blend of the affluent American Midwest and European culture, and this blend would affect his attitudes toward colonial expansion, the colonized of the empire, and his own view of himself.

During his time laying new wire Mitchell came under fire from "bandits" and was disgusted when his men found bodies of Spaniards—men, women, and children—killed by Cubans and thrown into water wells.[13] Firmly believing that Spanish control over Cuba had been best for the Cubans, he also expressed his faith that U.S. control would bring even more progress to the island. Cuban independence without American oversight would, he thought, be a disaster for the people of Cuba. His views were not unlike other colonial advocates. The British concept of the "White Mans' Burden" and the French Mission Civilisatrice also stressed the role of the European in bringing order and the benefits of European (and American) culture to the Third World, and these rationales were certainly employed to justify annexation in the late nineteenth century. Flushed by the recent victory over Spain, Mitchell could write glowing reports of American enterprises in Cuba, but he never seemed to be able to articulate just what needed to be done to ensure that what he saw as the blessings of

American-European civilization would be of direct benefit to the Cuban people.

Mitchell wrote to his father that February and March had brought an upsurge in Cuban activity against the American authorities. Again referring to Cuban resistors as "bandits," he wrote to his father about the nearly one hundred miles of telegraph wire he had just prepared to tie Havana, Santiago, and the interior together. In what must have been something of a surprise to John Lendrum Mitchell, William stated that he could not get a regular commission until 1901, and he believed that "it would be a good thing for me to stay in for five years anyway."[14] He wrote that he planned to return to Milwaukee for a leave, but he remained silent about either returning to Cuba or going to the Philippine Islands.

By this time William seemed determined to stay in the army until he could qualify for a Regular Army commission and appeared content to dedicate the next four years of his life to obtain the prize. While expressing typical colonialist views about Cuba, he had no illusions about the war on the island. He observed the dominant role that the U.S. sugar interests played in the coming and prosecution of the Spanish-American War, but he justified it by telling his mother, "The [Cuban] people cannot live without the sugar plantations because all the money that comes into the island comes from the plantations."[15] It was also clear that William wanted no part of the business world that spawned the vast sugar plantations in Cuba. He was drawing away from the business world, the source of Mitchell family wealth, and was moving toward an army career that offered what Mitchell at age twenty wanted most of all—action and recognition. The one fly in the ointment was, however, money. William had to rely on the continual generosity of his mother to supplement the low pay of an army lieutenant. Having already contributed funds for such things as a dress uniform, a sword, and horses, she also periodically sent money for William to continue his limited social life in Cuba. There can be little doubt that William Mitchell relished command and authority, and he exercised both quite well in Cuba.[16] Were action, recognition, command, and authority readily obtainable by a young man in his father's business and banking world in 1899? William Mitchell apparently did not think so.

In June Mitchell was informed that his reports on the condition of telegraph lines from Santiago to Havana had attracted the attention of the chief signal officer in Havana and would be sent to the chief signal officer of the army in Washington, D.C.[17] He took great pride in the fact that the vast majority of his superior officers were West Point graduates and that as a very junior officer he had been singled out for meritorious service in Cuba.[18] During the summer of 1899 he pressed his request for service in the Philippines with fervor. Mitchell found that he could put in a transfer on his own merits rather than asking his father to intervene for him with the War Department. Besides, William's parents had left the United States and were in Germany preparing to travel to France, so his father's intervention on his behalf was impossible at the moment. Mitchell would have to rely on the good opinions of his superiors and the judgment of the War Department in regard to his request for Philippine service.

And a good opinion of Lieutenant William Mitchell they did have. On June 14, Mitchell received a copy of a response from the army's chief signal officer in Washington to his chief subordinate in Havana that said in part, "As you [chief signal officer, Havana] state, it is a very creditable report, and indicates that this officer, despite his youth, is a man of ability, energy and intelligence. I have seen few reports giving so much information in clear cut form on a technical subject of such range."[19] Senior officers in Washington had to rely on reports from commanders in the field, and to have such recognition made a Regular Army commission and a transfer to the Philippine Islands only a matter of time. After the poor showing of the United States Army in the Spanish-American War, the military establishment in the War Department appreciated a junior officer who showed initiative, competence, and the ability to accomplish a mission with a minimum of supervision. Mitchell, however, realized that though he had accomplished much on his own, his father looked with disfavor on his going to the Philippines, where there was a state of armed insurrection against the newly established U.S. authority.

Mitchell felt it necessary, therefore, to explain his actions to his father. "I don't believe much in annexation," he wrote the senior Mitchell, "but we cannot get out of keeping the Philippines and preserve our dignity."[20] Truth be told, he certainly *did* think highly of American colonial acquisition in the Pacific, but still he could not bring himself to directly confront his father's well-publicized anticolonial views. At the end of July he wrote his father again: "As you probably know I have been ordered to the states probably for duty in the Philippines. Since you wrote to me, I made no effort for the assignment although I had applied for the place about three months ago when I returned from Santiago. . . . If I should get out there and you wish me to go to Europe next year I can resign there and take the steamer by way of the Suez canal."[21] Mitchell had indeed pushed for an assignment in the islands, and the idea of his going to Europe as a civilian had not entered the picture prior to several admonitory letters from his father. William continued to look for a glimmer of paternal approval for the course he had selected for himself. (Interestingly, William felt very comfortable in confiding his hopes and plans to his mother.)

By the end of July he had his assignment in the Philippines and planned a short leave in Milwaukee, or in Europe if his parents were settled there. He returned briefly to Washington to complete his degree require- ments at Columbian, which would be important in the Regular Army with its coterie of West Point graduates. His father had earlier raised the ques- tion of joining the army prior to receiving his degree. He tried again to express his feelings to his father about the assignment:

> I hope that you will not oppose my going [to the Philippines] as the only reason I came into the army was to get some active service and if I do not get it [I] will resign as there is very little in it for me. . . . It has been a very good thing I think and have benefitted by it more than a years work in college I think and capping it off by a trip to the Philippines and getting some good active service would be a fine thing. I do not think that you need have any doubts about my keeping well as I have had a pretty good test in a good many ways. Now that my order is out and if I did not go it would look as I had "gotten cold feet" as we say. I guess that you can imagine how anxious I am to go.[22]

He did not convert his father to his Philippine adventure but did return to Milwaukee for leave before undertaking the long and arduous journey across the Pacific Ocean. While there he had the opportunity to see Douglas A. MacArthur, the long-time family acquaintance.

Douglas MacArthur and his mother had moved to Milwaukee from an army post in Texas in 1897. Arthur MacArthur was at that time serving on the staff of the Department of the Dakotas, headquartered in St. Paul, Minnesota. Douglas's mother felt that Milwaukee offered better education- al opportunities for her son, and there were old family friends in the city. The young MacArthur spent his time preparing for entrance into West Point. While in Milwaukee MacArthur was introduced to William Mitchell's sister Harriet, and they had a brief romance. The two young men met socially during William's leave but did not strike up a close association.[23] William was a veteran of Cuba, a handsome young man in dress uniform, and was on his way to the Philippines. Douglas had just completed his first term at West Point and was no match for the glamorous Mitchell who had such a distinguished service record in Cuba. Their paths would not cross again until two decades later, during the Great War in France.

Mitchell was going to a Philippines mired in confusion and an insur- rection against U.S. authority. When the Spanish-American War broke out, Commodore George Dewey landed Emilio Aguinaldo y Famy in Luzon, ostensibly to fight against the Spanish. Aguinaldo had been a prominent member of the Filipino revolutionary movement but since 1897 had been exiled in Hong Kong. No sooner had Aguinaldo landed than he proclaimed the Republic of the Philippines, with himself as president with near dictatorial powers. With the surrender of Spanish authority, clashes between the American and Philippine troops grew in number.[24] By the time Mitchell received his final orders to report to the islands, there was open warfare between the two sides. Unlike Cuba, where dangers were restricted to small firefights between U.S. troops and so-called bandits, the Philippines offered actual combat operations on an ongoing basis. Very quickly Mitchell became involved in fighting against Aguinaldo's forces. He described the insurgents as "miscreants" and recounted stories of atrocities against U.S. forces.[25] These outrages were for Lieutenant Mitchell a clear sign that the Filipinos were as much in need of America's civilizing and progressive influence as the Cubans had been.

After a brief orientation, Mitchell was assigned to duties in Manila, where he commanded a signal company. He liked the growing American social life in Manila as the wives and daughters of American officers and administrators arrived in the capital city. Mitchell expressed his expansionist tendencies to his mother, writing that he believed that the United States would pacify the country in a few years at best and then would provide stability and progress for the islands.[26] During his first months in the islands he continued to formulate his views on the benefits of American imperialism. He wrote to his father, "These people are a fairly practical sort and when they see that we mean them no harm, that they will be tied down to no secular religious zone and have the right of sufferage [t]hey will rapidly discontinue this disorder. Of course by instinct some of them like to roam around and do a little scrapping but they will have to stop it now."[27] Beyond that, William told his skeptical father little about his activities in the Signal Corps.

Mitchell did basically the same type of work he had done in Cuba, except that the Philippine Islands were much larger and the telegraph system much more complex. William found that he preferred to be in the field with his troops rather than in Manila with the staff. Despite the social life in the capital, he relished the work and, being in excellent physical condition from sports and horseback riding, was called upon for difficult missions into the interior. He told his father that he had been under fire several times. The signal unit under Mitchell was protected by a detail from the 17th Infantry Regiment, but Mitchell, as an officer, participated in various firefights.[28] John Lendrum and Harriet Mitchell were now settled in Europe, with the children enrolled in schools in Germany and in France. William planned to apply for a leave once General Arthur MacArthur took command in Manila. He outlined an ambitious itinerary through India, the Suez Canal, Greece, and North Africa to get to France.[29]

Mitchell realized that his time in Cuba was a great advantage for him because senior officers he knew there were now serving in the Philippines. It was fairly clear to William that the army was a small organization, and his reputation as an officer who could get things done now served him well. He made himself known to General Arthur MacArthur, whom he had met through his father in Milwaukee. William Mitchell was learning to play the "old army game," and he was playing it quite well, in fact. Low pay still bothered Mitchell, and he asked his mother to send him a Mauser pistol when she and his father visited Germany.[30] After eight months of service in the islands Mitchell was still pondering his future. He was twenty-one years old and could apply for a Regular Army commission—or he could leave the army. This ambiguity was certainly not unusual for a young man of his age. The excitement of action continued to intrigue Mitchell, and he was given more infantry and cavalry to use as he worked with the telegraph system. His superior officers were very pleased with Mitchell's ability to relate to Filipinos and to gather vital information. Despite his wavering attitude toward an army career, he took umbrage at newspaper reports that the soldiers and their officers in the Philippines were drinking a great deal and that soldiers were lying on the streets in a drunken stupor. "[W]hat counts in the army," he informed his mother, ". . . is the result of attention to duty and application of details in nearly all cases."[31] There could be no question that Lieutenant Mitchell had proven that to his commanders.

In the spring of 1900 Mitchell would complete almost two years of overseas service, and he was promised an assignment to the newly established Signal School at Fort Myer, Virginia. He had earned the chance to further his military career, and he made the choice to go to Virginia rather than resign. After a visit with his parents in Europe, Lieutenant William Mitchell began his training in the United States. Years of action and adventure abroad, however, left him dissatisfied with schooling and post duties in the United States. For a while he again considered leaving the army. There was a possibility, Mitchell believed, that with his considerable language capabilities he might be able to find a position as a military attaché somewhere in Europe for three or four years before leaving the army, if indeed he planned to return to civilian life.[32]

Another factor in Mitchell's life made him think seriously about his future plans. With his parents in Germany, Mitchell visited old friends of his mother's, including the Stoddard family of Rochester, New York. There he met Caroline Stoddard, who had just finished her junior year at Vassar College. It was a pleasant time for Mitchell, and Mrs. Stoddard regaled him with stories about herself and his mother when they were young girls. His attention was focused on young Caroline, who was intelligent, athletic, and determined to finish her senior year at Vassar. "She is the finest girl in every way," he wrote to his mother. He hoped to visit the Stoddards late in the summer, and it was obvious that William was romantically interested in Caroline. While pursuing signal training at Fort Myer, he seemed distracted by the young woman in Rochester.[33] Caroline promised to visit him at Fort Myer in the fall prior to her return to Vassar.[34]

Caroline would have to wait to visit William, however. He had been given the opportunity of a transfer to Alaska to do signal work. On August 4, 1901, Mitchell received orders to report to Fort Egbert, Alaska, the following month. Major General Adolphus Greely, the chief signal officer of the United States Army, who was aware of the outstanding record of Lieutenant Mitchell in Cuba and in the Philippines, felt that Mitchell's industriousness and physical condition would make him an excellent choice for what would be a difficult mission.[35] Greely, who would be a lifelong role model to Mitchell, had a quick, inquiring mind, and he would be one of the first general officers to understand the potential of airpower. Greely showed great interest in Mitchell's career and did more to direct the youthful officer than did his father. It was Greely who cemented Mitchell's commitment to the army. Anxious to get to Alaska prior to the onset of winter, William Mitchell was hopeful that a successful tour in Alaska would secure the rank of captain for him. With General Greely's support he would become the youngest captain in the U.S. Army.

The laying of military telegraph lines in Alaska had been mandated by Congress in May 1900 in order to carry civilian and military traffic from various parts of the vast territory. The Alaskan gold rush of the late nineteenth century had brought great numbers of prospectors and businessmen into Alaska, and it was now vital to establish regulated communication with the territorial capital of Juneau and major towns. The Washington-Alaska Military Cable and Telegraph System was instituted, and those who served with it were known as WAMCATS.[36]

Mitchell knew that his time in Alaska would be hard due to the winter weather and the lack of transportation. He had a very short time to prepare for the tour in Alaska, which promised to be about two years in length. Before he left, Mitchell visited Milwaukee, and in early August he went to see Caroline in Rochester, at which time he and Caroline informed Dr. and Mrs. Stoddard of their engagement. This did not please Mrs. Stoddard, who made William and Caroline pledge that the engagement would not be made public until she graduated from Vassar at the end of the academic term in 1902. William promised his mother that they would make no wedding plans until the families could meet and discuss them fully. Mitchell also told his mother that after the Alaska tour he would apply for a military attaché post

and then return to Milwaukee and the family business. Evidently the
Stoddards were concerned over the low pay of an army officer and what
impact that might have on their daughter. But William was committed to
the marriage when he told his mother, "As you know Caroline is the only
girl that I have cared much about and I probably wouldn't ever care about
anyone else in the same way, not for a long time at least."[37] At any rate,
both William and Caroline would have time to ponder the steps they had
taken toward marriage.

William Mitchell arrived in Alaska during the winter of 1901, and he
found Alaska to be more uninviting than he had imagined, with winter temperatures that dipped to fifty or sixty degrees below zero. It was difficult to
find horses in the territory, and he soon realized that much of the trip to the
interior would have to be by dog sled. He served at Forts Egbert and St.
Michael until late November 1901, and then new orders sent him in the
dead of winter to Skagway to prepare to lay telegraph lines.[38] As part of the
mission, Mitchell faced the daunting task of moving several hundred tons
of supplies to complete 420 miles of telegraph lines. General Greely
remained pleased with Mitchell's work and comprehensive reports.[39] His
physical condition and the good reports by his superiors lifted Mitchell's
spirits, and he wrote his mother, "I have had a very nice little stay here in
Skagway and will enjoy this Alaskan service more than any foreign service
I could have."[40]

Time was at a premium for Mitchell during the winter of 1901–1902,
but he was very productive. When spring came he marveled at the wild-
flowers and trees, but he found that mosquitoes were worse in Alaska than
they had been in Wisconsin, Cuba, or the Philippines. His romantic interest
in Caroline Stoddard had not diminished by 1902, and he knew that in May
she would graduate from Vassar and that plans for their wedding could be
made public. He was certain that in a year he would be promoted to the
rank of captain because of the support of his superiors in Alaska and
General Greely in Washington. Perhaps that would show the Stoddards that
his career was on the rise. He had altered his position again about leaving
the army and told his mother that the only way he would consider resignation would be if the army ordered him to the Philippine Islands after his
Alaska tour of duty.[41]

Mitchell fully expected to leave Alaska during the summer of 1903, but
gold was discovered near the town of Fairbanks and settlers and prospectors poured into the area. Fairbanks was in the interior and needed
increased telegraphic communication with the major towns on the coast. It
fell to Mitchell to undertake the mission of raising poles and laying wire for
several hundred kilometers over wild and wooded terrain. The 1902 gold

discovery at Fairbanks pushed Mitchell to complete the lines, which he did in order to keep his 1903 departure date intact. When he left Alaska in July 1903 he had earned a special place among Signal Corps officers as a man who could tackle the most difficult jobs and accomplish missions on time. General Greely assured Mitchell that he would wear the two silver bars of a captain as soon as he returned to the United States. At the age of twenty-four, William Mitchell became the youngest captain in the Regular United States Army.

By the time Mitchell returned to the United States his parents had returned to Milwaukee. Though John Mitchell continued to work in various banking concerns, his health was frail. Harriet was left to battle the problems caused by declining family fortunes. It was she who kept the beloved Meadowmere in the Mitchell family, raising and selling her esteemed horses and devoting part of the estate to crops. William's younger brother, John L. Mitchell Jr., seemed well on his way to becoming the agriculturist of the family.[42] John Jr. also attended Racine College and then elected to attend the growing University of Wisconsin at Madison. The senior John Lendrum Mitchell had helped found the agricultural college of the university, and he had passed on to his namesake a love for the land and for Meadowmere. William Mitchell, however, had more on his mind than horses and crops. The Stoddards had finally agreed to a December wedding of their daughter to the accomplished and handsome young captain.

Mitchell traveled to Washington, where he was warmly received by General Greely and Major George P. Scriven, who would become a major force in the development of military aviation. Captain William Mitchell had two major concerns: first, to arrange a leave in order to marry and have a honeymoon; and second, to find out the location of his new assignment. He certainly did not want an overseas assignment, which would separate him from his new wife. Greely informed Mitchell that he would be assigned to a signal unit at Fort Leavenworth, Kansas, and with that settled Mitchell went to see Secretary of State John Hay to ask for visas to visit Mexico during his upcoming honeymoon. With leave, assignment, and visas taken care of he left for Rochester to prepare for the wedding.[43]

Caroline Stoddard was a fine match for William Mitchell. She came from a highly educated family and prepared for Vassar by attending the Crittenden School in Rochester as well as various French schools. She did well in her academic studies at Vassar, acted in several plays, and was elected president of the drama society in 1902. Caroline was also an active member of the Athletic Association, the Marshall Club, and the Rochester Club; served on the Seniors to Sophomores Committee; and was vice president of the Dickens Club, a literary organization. Her picture in the 1902

Vasarion shows her to be an athletic, attractive brunette. A member of the Episcopal Church, Caroline was socially adept and active, but she also displayed an independent and daring personality, a good complement to William Mitchell. Caroline's father, Enoch, a graduate of Yale University, encouraged her to pursue ambitious interests that other women at that time usually did not.[44]

The marriage of William and Caroline promised to be a major social event. William would cut a dashing figure at the wedding in his blue uniform with braid, sporting two medals: the Spanish-American War medal and the Philippine Army Campaign medal. His mother planned to arrive several days before the ceremony to assist the Stoddards with the preparations. His father seemed almost detached from the preparations, and there were questions as to his attendance because of his health. Money for the planned lengthy honeymoon was tight, and William spent a good deal of time finding the cheapest way to travel to Tampa, Florida, from Rochester. This turned out to be a circuitous route by train west of the Mississippi, approaching the port of Tampa from the west.[45] From Tampa the newlyweds would go to Cuba and then on to Mexico.

On December 2, 1903, the couple were married and left for their honeymoon. After a few days in Cuba, where Mitchell saw many of his old Spanish-American War comrades, William and Caroline departed for Mexico City. They resided at the Hotel Reforma in a fashionable district of the city. Like his father's trips to Europe with the children, Mitchell's time in Mexico was spent viewing ruins and Mexican-American War battlefields. In a letter to his ailing father, Mitchell described the ruins and the historical significance of Aztec civilization. He also complained about the expense of the trip. Indeed, no matter how enthusiastic William Mitchell was in his letters concerning historical sites and Mexico in general, money worries crept into his communications with his mother and father. He and his new wife would continue their trip, he wrote them, "if the money holds out." He requested that the army send his pay for November and December 1903 to him in Mexico.[46]

There can be no question that William and Caroline were very much in love, and, more important, that they shared many interests. In Mexico they explored the precolonial and colonial history of the country; they also reveled in riding horses, trekking about the countryside, and hunting. Caroline wrote often to her mother-in-law, and her letters reflect true happiness. She expected that very soon her husband would be given orders for the Philippines, and she was looking forward to the experience. She was quick to inform Harriet Mitchell that they were also enjoying what William called "the society act," and that many of their conversations at social gatherings were carried on in French, which she spoke with facility.[47] The honey-mooners traveled to Chapala, in the state of Jalisco, to spend several weeks hunting and riding horses with friends.[48] What Caroline did not state in her letters home, or possibly did not know, was that the stay in Jalisco stretched their scant funds much more than remaining in Mexico City would have.

It is clear that Caroline also fully expected to enjoy her life at Fort Leavenworth, Kansas, where her captain-husband would have a command and would have a chance to attend the very important new schools established there. Fort Leavenworth was no longer considered the frontier and was close to the growing urban area of Kansas City. On bluffs above the Missouri River, Leavenworth was an old post with comfortable housing, an active social life, and many possibilities for hunting and horseback riding. Because Washington was so expensive for a young officer and the California posts were so far from Wisconsin and New York, there could not be a better assignment for Captain Mitchell and Caroline.

The Mitchells had just settled into their quarters at Fort Leavenworth when word arrived that John Lendrum Mitchell had died in Milwaukee on June 29, 1904. William had mixed emotions over his father's death. The distant father had compiled an impressive record of achievement in Wisconsin and Washington and was a very difficult model to copy. In addition to his well-known political views and efforts as a senator, he bought the city of Milwaukee its first steam-powered drawbridge. As an educational reformer he purchased with his own funds textbooks for children whose parents did not have the means to buy them. A true scholar, he helped to maintain over thirty scholarships at the University of Wisconsin.[49] One Mitchell biographer has written, "If he [William Mitchell] were to match the record of his grandfather and father, it was in the army that his reputation had to be made."[50] After 1904 William Mitchell rarely spoke of his illustrious father.

With John Lendrum Mitchell's death, Harriet Danforth Becker Mitchell became the anchor for the family. William's letters would change from "My Dear Mother" to "My Dear Mummy," rather strange for an army captain approaching his thirtieth birthday. Mitchell was now the captain of his own future, but steering the proper course through the old army would be rife with challenges. Even so, Mitchell was prepared. In addition to the patronage of Major General Greely, Mitchell had important political contacts in the U.S. Senate and the Wisconsin Democratic Party. He was always ready to use that political leverage, to play on the Mitchell legacy, when it suited him. The youngest captain in the army was bound to create jealousy, and William Mitchell was not one to hide his intelligence, his record, or his political friends under a basket. The next decades would take him even higher and into areas of controversy as the United States faced its first great foreign crusade.

NOTES

1. Mitchell to Mother, 2 January 1899, in the William Mitchell Papers, Library of Congress, Washington, DC, Carton 19A. (Hereafter cited as Mitchell Papers).

2. Ibid.

3. *Congressional Record, Proceedings and Debates of the Fifty-Fifth Congress, Second Session,* 31 (Washington: Government Printing Office, 1898), 6188.

4. Mitchell to Father, 3 January 1899, Mitchell Papers, Carton 19A.

5. Ibid.

6. Mitchell to Father, 14 January 1899, Mitchell Papers, Carton 19A.

7. Ibid.

8. U.S. Congress of the United States, *Biographical Directory of the United States Congress* (Washington: Government Printing Office, 1989), 1514.

9. Colonel H. C. Dunwoody, Chief Signal Officer, VII Corps, to Mitchell, 2 February 1899, Mitchell Papers, Carton 19A.

10. Mitchell to Dunwoody, 11 February 1899, Mitchell Papers, Carton 19A.

11. Mitchell to Father, 14 February 1899, Mitchell Papers, Carton 19A.

12. Ibid.

13. Mitchell to Mother, 28 March 1899, Mitchell Papers, Carton 19A.

14. Mitchell to Father, 30 March 1899, Mitchell Papers, Carton 19A.

15. Mitchell to Mother, 6 April 1899, Mitchell Papers, Carton 19A.

16. Ibid., 14 April 1899, Mitchell Papers, Carton 19A.

17. Office of the Chief Signal Officer, Division of Cuba, to Mitchell, 1 June 1899, Mitchell Papers, Carton 19A.

18. Mitchell to Mother, 9 June 1899, Mitchell Papers, Carton 19A.

19. Office of the Chief Signal Office, Division of Cuba, to Mitchell, 14 June 1899, Mitchell Papers, Carton 19A.

20. Mitchell to Father, 13 July 1899, Mitchell Papers, Carton 19A.

21. Ibid., 23 July 1899, Mitchell Papers, Carton 19A.

22. Ibid., 29 July 1899, Mitchell Papers, Carton 19A.

23. D. Clayton James, *The Years of MacArthur, Volume 1, 1880–1941* (Boston: Houghton Mifflin Co., 1970), 64–65.

24. For a brief but coherent account of U.S. actions in the Philippines, see Albert A. Nofi, *The Spanish-American War, 1898* (Conshohocken, PA: Combined Books, 1996).

25. Mitchell to Father, 12 January 1900, Mitchell Papers, Carton 19B.

26. Mitchell to Mother, 25 January 1900, Mitchell Papers, Carton 19B.

27. Mitchell to Father, 28 January 1900, Mitchell Papers, Carton 19B.

28. Ibid., 10 March 1900, Mitchell Papers, Carton 19B.

29. Ibid.

30. Mitchell to Mother, 16 April 1900, Mitchell Papers, Carton 19B.

31. Ibid., 1 June 1900, Mitchell Papers, Carton 19B.

32. Ibid., 1 July 1901, Mitchell Papers, Carton 19B.

33. Ibid., 14 July 1901, Mitchell Papers, Carton 19B.

34. Ibid., 30 June 1901, Mitchell Papers, Carton 19B.

35. Alfred F. Hurley, *Billy Mitchell: Crusader for Air Power,* 2nd ed. (Bloomington: Indiana University Press, 1975), 9.

36. *Alaska Communications System, 49th Anniversary, 1900–1949* (Alaska: ACS, 1949), 3.

37. Mitchell to Mother, 19 September 1901, Mitchell Papers, Carton 19B. (This letter was sent from Seattle, Washington.)

38. *Alaska Communications System,* 3.

39. Mitchell to Mother, 15 December 1901, Mitchell Papers, Carton 19B.

40. Ibid.

41. Ibid., 23 June 1903, Mitchell Papers, Carton 19B.

42. E. M. Kelly, "The Man Who Wouldn't Shut Up," *Colliers* 76 (December 12, 1925), 12.

43. Mitchell to Mother, ca. 23 November 1903, Mitchell Papers, Carton 19B.

44. Information provided by the Alumnae and Alumni Vassar College Association, Vassar College, Poughkeepsie, New York.

45. Mitchell to Mother, ca. 24 November 1903, Mitchell Papers, Carton 19B.

46. Mitchell to Father, 7 January 1904, Mitchell Papers, Carton 19B.

47. Caroline Stoddard Mitchell to Her Mother-in-Law, 10 January 1904, Mitchell Papers, Carton 19B.

48. Ibid., 25 January 1904, Mitchell Papers, Carton 19B.

49. Kelly, "The Man Who Wouldn't Shut Up," 12.

50. Hurley, *Billy Mitchell,* 16.

THREE

THE GREAT WAR, 1917

A FTER THE BURIAL OF HIS FATHER, WILLIAM MITCHELL
returned to Fort Leavenworth to resume his duties as a Signal Corps
company commander. Fort Leavenworth was considered "the intellectual
center of the army" at the turn of the century, and the Signal Corps was
thought to be an up-and-coming element of the army due to the developing
rapid communication of the telegraph. First used in the Civil War, and later
to link far-flung frontier posts, the telegraph was now indispensable for
military operations, and Mitchell had built a reputation on laying lines. The
invention of the telephone by Alexander Graham Bell, who first displayed
his newly patented telephone at the great 1876 Exhibition, offered to for-
ward-thinking signal officers the possibility of instant voice communications between commanders and their units on the battlefield.

In the 1880s the army had opened a School of the Line in Leavenworth, and plans were underway for an expanded General Staff School and
a Signal School. The only two other places where William Mitchell's
advancement could be assured would be Washington and the Philippine
Islands. Mitchell always had ambition, but at Leavenworth, with friends in
Washington, he could pursue his career in pleasant and inexpensive surroundings. Mitchell took no notice, however, of a momentous event on a
sandy North Carolina beach, free of trees and scrub, on December 17,
1903—when Wilbur Wright made a flight of 120 feet in the air with a
machine-powered craft. The Kitty Hawk flight would have dramatic consequences for the entire world, and it fired the imagination of young army
and navy officers, who saw the possibilities of a new battlefield defined by
width, depth, and now height. The new century promised to be an exciting
time for young and ambitious officers like William Mitchell.

Captain Mitchell very quickly attracted the attention of his superior officers, and they marked him as an officer on the rise. He was selected to attend the demanding School of the Line, a requirement for promotion, and graduated as the school's Distinguished Graduate. The next step would be to attend the Army's Staff College, but that was some time in the future. In addition to outstanding performance in the classroom and in the field with his signal company, Mitchell became a father. On August 18, 1906, Elizabeth Mitchell was born at Fort Leavenworth. Both Harriet Mitchell and her longtime friend Caroline Butts Stoddard were pleased to have a grandchild, and Captain Mitchell spent much time with his daughter. He was sincerely happy with his home life and with the addition to the family. As soon as the young Elizabeth was able, William and Caroline began teaching her to ride horses.

While at Leavenworth Mitchell worked with the head of the signal school, Major George O. Squier, and he would prove to be a very valuable contact for the young captain. Squier was interested in the military use of the balloon and in what was called "aerostation." For someone with the personality and background of William Mitchell the air held a certain fascination. While serving in Alaska in 1903, Mitchell had used a box kite and air currents to go into the air, to an altitude of 100 feet, to study the terrain so that he could find the proper locations to place his telegraph poles.[1] Major Squier would later play a pivotal role in the development of airpower in the Great War, and he would also be influential in Mitchell's career in aviation.

Life for the Mitchells at Leavenworth was much like that of any other officer, except that it was clear Captain William Mitchell was marked for rapid advancement. In 1908 Mitchell entered the Army's Staff College, and in July 1909 he earned the honor of once again being named the Distinguished Graduate. This guaranteed him a prize assignment as his time at Leavenworth came to an end: he was to become the chief signal officer for the Department of the Philippines. On November 14, 1909, his second daughter, named Harriet after his favorite sister, was born.

One of the tools that the army used to measure the progress (or lack thereof) of its officers was the annual efficiency report, written by a commander and sent to Washington. Mitchell's performance as reported by his commanding officers continued to be exemplary. At Leavenworth in 1905, he was described as "an earnest, zealous, efficient officer; he is intensely interested in his professional work and in the appearance of his company; an enthusiast in whatever he undertakes." During the terrible carnage of the San Francisco earthquake of 1906, Mitchell was chief of the army's First Relief Station, and his commander wrote that in the midst of such misery Mitchell's work demanded, "physical endurance, the exercise of tact and discretion and judgement of the highest order." In 1908, however, Mitchell's report cited his first bout with rheumatism, which kept him from classes at Leavenworth's School of the Line. But he made up his work and was retained for the Staff College course in 1909. What these reports told the staff was that Mitchell was first-rate, an officer destined for higher assignment.[2]

From 1910 to late 1911, Mitchell and family followed the course that hundreds of other army officers had followed, with nothing spectacular to mark those years in the islands. As he had in the past, Mitchell compiled a record of success as chief signal officer and as commander of Company L, Signal Corps, at Fort McKinley. In 1912, he received orders to report to Fort Russell, in Wyoming, to assume command of Company I, Signal Corps. Mitchell had been told that he was under consideration for a post in the General Staff in Washington, D.C., and with that choice appointment would come the gold leaves of a major. Reporting to his new assignment in Wyoming, Mitchell had to assume the life of a bachelor officer: Caroline took the two children back to Rochester, for she did not want to settle into rustic quarters at Fort Russell when, in a short time, she would be looking for a suitable home in the nation's capital.

Mitchell was excited about this obvious boost to his career. He told his mother that he had been officially notified he would become a member of the General Staff on March 5, 1912. William Mitchell claimed that the assignment had come as a total surprise, which it certainly had not, and then went on to give her a detailed description of his duties. He was thirty-one years old and would be the youngest General Staff officer in the short history of that institution.[3] The chief of staff at that time was Major General Leonard Wood, himself something of an oddity in the army. Born in 1860 in Massachusetts, Wood first embarked on a medical career, receiving his degree in 1884 from Boston City Hospital. Rather than open a practice in Boston, Wood became a contract surgeon for the army. While serving in the West he was awarded the Medal of Honor, and he earned the star of a brigadier general as a combat commander in Cuba in 1898. From 1903 to 1908 he served in the Philippines.[4]

In the Philippines Mitchell had served under Major General James Franklin Bell, himself a former chief of staff of the army, and Mitchell was certain that Bell and others brought him to Wood's attention. It did not hurt William Mitchell that he was close to the Adjutant General of the Army, Major General Fred C. Ainsworth. Mitchell told his mother that he counted on Ainsworth to pave the way for his transition to the General Staff. Upon his departure from his last assignment in the Philippines, Mitchell had been ordered to look into military conditions in Manchuria after the Russo-Japanese War of 1905. His coherent and cogent report was important to the army and helped Wood decide Mitchell would come to Washington.[5]

The Mitchells would be something of a strange sight in the War Department and the General Staff. When Mitchell went to Washington he was thirty-two years old; Caroline was thirty-one years old, considerably younger than the circle of army wives. George O. Squier, an old friend who would have been close to the Mitchells, had received orders to assume the duties of military attaché in London. However, problems in Cuba and along the U.S.-Mexican border intervened, and Mitchell was ordered with his company to Fort Bliss at El Paso, Texas, to prepare telegraph lines along the difficult border. After a few months on the border he relinquished command of Company I and proceeded to Washington. Mitchell was exultant over the assignment, to be sure, but he also looked forward to having a stable environment for his family. In letters to his mother he carefully described how well Elizabeth was reading and how quick Harriet was learning. He looked forward to enrolling little Harriet in one of the capital's good public schools, and he told his mother that certainly they would attend Sunday school at a local Episcopal church. Caroline had found a nice house in the neighborhood of I and 17th Streets, where a large number of officers employed at the War Department and General Staff resided.[6] All in all, the move to Washington and the pending promotion to major seemed like a dream come true for William and Caroline Mitchell.

The General Staff was only a decade old when Mitchell arrived in Washington. The poor showing of the army in the Spanish-American War had prompted President William McKinley to bring Elihu Root, a brilliant New York lawyer, to the nation's capital as Secretary of War. Root's mandate was simple: reform the army and modernize the institutions that had failed so badly in 1898. Root went to work immediately and in a few months had a series of suggested reforms for the army, including the establishment of a War College, merit selection of officers, and large-unit training. In 1903 Root proposed to Congress the crowning achievement of his secretaryship, the establishment of the General Staff. The old title of commanding general would be changed to chief of staff of the army, indicating that the new position would manage the everyday affairs of the army and plan for future contingencies with the assistance of a trained staff. Education and merit became the cornerstone of the new staff.[7] It was, then, no accident that a young and accomplished officer like Mitchell would be detailed to the General Staff's signal section.

It was an exciting time for Mitchell to be on the staff. Mexico was in the midst of a revolution, and the United States was very concerned about security and stability along the border. Mitchell had experience in Mexico, since he had honeymooned there and had very recently prepared telegraph lines in the Fort Bliss–El Paso, Texas, area. The Milwaukee Loyal Legion had requested that Mitchell return to his native state to give an address on the army and the present situation along the U.S.-Mexican border, which he agreed to do. There was nothing unusual about a knowledgeable officer speaking to a group that usually supported American military strength, but there was another reason for Mitchell to return to Wisconsin. He had been told that if hostilities broke out between Mexico and the United States, the adjutant general and governor of the state considered Mitchell the ideal candidate to command a Wisconsin volunteer regiment. Despite widespread army skepticism about the viability of the volunteer militia, Mitchell maintained close ties with the state organizations. Major General John O'Ryan of the New York National Guard had specifically asked for Mitchell to accompany him as a representative of the General Staff at a review of the New York division. Mitchell was learning to play a political game, and to play it quite well.[8]

The first Christmas season in Washington for the Mitchells was a happy one. In the living room was a fireplace where the children had the traditional stockings, and there were presents from the grandparents.[9] This pleasant holiday would be the last one for some time to come, for clouds were on the horizon for William and Caroline Mitchell. They had concerns over money and over Mitchell's health—specifically, a return of the eye problems that had plagued him while at Racine College. In April, a local doctor prescribed belladonna for his reoccurring problems, and for several weeks he had to remain at home. The cost of living in Washington was high, straining Mitchell's modest income. Out of his modest captain's salary came payment for the boarding of horses, dancing and piano lessons for his daughters, and the various social costs of being in the capital city. Mitchell's mother regularly sent him money to pay various bills, but that did not cover all of the expenses. "I want to take a thorough course in fencing . . ., and if practicable would like to belong to the Chevy Chase Club, as the tennis, golf, etc. is good in addition to the fact that most people here belong to it," he wrote to his mother. He then asked if his mother could send him $2,000 to help defray various expenses.[10]

Captain William Mitchell continued his duties on the General Staff as conditions with Mexico deteriorated. In late April 1914 Governor Francis S. McGovern wrote to President Woodrow Wilson to offer the National Guard of Wisconsin for service if the situation required it. He also informed Wilson that the state would raise a volunteer cavalry regiment for service and requested that Mitchell be allowed to return to Wisconsin to raise the unit.[11] This did not come as a surprise to Mitchell because there had been hints for several months. He had encouraged the interest in his command of a regiment that, of course, would have carried with it a colonelcy in the Wisconsin National Guard. It was a dangerous game for the captain to play because of the attitude of the Regular Army toward the militia, and Mitchell had a choice assignment to the General Staff, a post many Regular Army officers wanted badly. The question of Mitchell's returning to Wisconsin was overshadowed by the outbreak of a general war in Europe in August 1914. Though the General Staff understood the president's orders of neutrality, there were many lessons to be learned from the great conflict taking place on two fronts in Europe. The General Staff's full attention would be focused on the severe battles in France and in East Prussia from August into the fall of 1914.

The General Staff was busy monitoring the fighting in Europe, trying to make sense of the cables they received from military attachés in European capitals. Mitchell was at the center of the activity, and he took advantage of his position and the incoming messages to write for the *Chicago Tribune,* identified only as "military expert." This was against War Department policy, but to earn extra money Captain Mitchell was willing to violate the standing directives of his service.[12] Mitchell's articles reflected the gloomy predictions of the military attachés; he personally believed that by the first week of September 1914 France would be decisively defeated by Germany. He wrote his mother that in the future the United States might have to fight Germany and that America would probably have to face Japan as well. He posited that the competition between the United States and the Japanese empire in the Pacific was bound to bring on conflict in the near future.

With Mitchell working about fourteen hours a day, often seven days a week because of the war, he and Caroline decided that she and the children should vacation without him. They decided to begin their time of relaxation after the first of September because the rates in Atlantic City, New Jersey, would be lower then. While she was gone, Mitchell requested that he be sent to Europe as an observer as soon as possible because he believed that the actual combat would be over soon.[13] William and Caroline had discussed this possibility, and she was aware of her husband's request before she and the children left for their vacation. During September, however, word reached Washington of General "Papa" Joffre's counterattack against the Germans on the Marne River, which halted their advance into France. The Miracle of Marne reversed France's military fortunes and ensured that the war would indeed continue. The urgency to get to Europe faded, and Mitchell resigned himself to continue on with his staff duties.

By December the Mitchells were again in dire financial straits, and Mitchell's mother responded to his pleas for help with a substantial gift. Mitchell was writing for newspapers as much as possible, continuing to violate military policies. In December 1914 Mitchell seriously considered an offer from a South American company to act as an agent in an importing scheme. The company, with assistance from their minister in Washington, would negotiate a loan in the United States for $4 million at 6 percent interest. They would in turn ship animal hides to the United States at a rate of nearly $2 million per year. If Mitchell helped to negotiate a loan and repayment in hides, he would receive what he said were thousands of dollars in commissions. Mitchell considered a firm in Milwaukee and asked his mother to sound out the company to see what interest they might have in the deal. Making it clear to his mother that the United States government would be involved, he asked her to give him her advice.[14] Nothing came of the scheme, but it does reveal that Mitchell evidently thought little about a conflict of interest representing a foreign concern in dealings with an American company in Milwaukee. Captain Mitchell appeared to believe that he could serve two masters—money and the United States Army.

Mitchell was part of an evolving Signal Section of the General Staff. The war in Europe brought into focus many changes in the way nations fought their wars, and aviation attracted much attention. In 1913 and again in 1914 Congress addressed the structure of aviation by creating a section devoted to the new way of waging war above the battlefield. Much of this growing awareness of flight was due to the complete and coherent reports sent back to Washington by Major Squier, military attaché in London, who had ample opportunity to observe modern combat and became increasingly fascinated by the airplane. In late 1914 Squier became convinced that air observation offered a considerable boon to the army that employed it properly. Because surprise, a major principle of war, was critical on the battlefield, the force that could detect by air the preparations and movements of the enemy had a decided advantage.[15] Squier returned often to the Western Front, and he realized that the old horse cavalry, once the eyes of the army, could not function on the modern battlefield. Emerging trench warfare, the machine gun, and dramatically improved indirect fire artillery rendered the horse soldier obsolete. But no army can survive long without accurate and timely reconnaissance, information, and observation. The airplane had to replace the horse, which on the World War I battlefield could no longer maintain its traditional role.

Squier was a close observer of what the British and French were doing in aviation and in establishing depots to support the maintenance and repair parts needs of aircraft. In early December, while visiting the Western Front,

Major Squier experienced air bombardment for the first time. On December 20, 1915, at the invitation of the French, he toured the front and watched an impressive use of ground artillery and the operation of French antiaircraft artillery against German planes. There was a massive amount of firepower directed against the Germans, with little effect.[16] In a report to Washington, Squier wrote, "Without indulging in prophecy, nothing is more certain than that the era of war in the air is upon us."[17]

Mitchell read the Squier reports, but between 1914 and 1915 he manifested only slight interest in aviation. The army had the 1st Aero Squadron stationed in Texas, and oversight of the squadron fell to the signal section of the General Staff. In 1914 an Aviation Section of the Signal Section of the General Staff was created, and Lieutenant Colonel Samuel Reber became its first chief. Reber, who had a difficult personality, had been chief of the aeronautical division and therefore had a good deal of knowledge about air matters. Despite general orders establishing a flying school at San Diego, California, there was little actual training or flying. American aviation was at a standstill, while combat flying developed at a rapid pace in Europe.[18]

Mitchell handled the reports coming from Europe, and though he realized the importance of air operations, he did not pursue the possibility of earning his wings. There were airpower advocates, civilian and military, who even viewed William Mitchell as antagonistic to the development of American air. Mitchell did not manifest any interest in joining the Aero Club of Washington. The club boasted among its members those individuals who were becoming influential in air matters, including Alexander Graham Bell, Glenn Curtiss, Lieutenant Benjamin Foulois, Lieutenant Frank P. Lahm, George O. Squier, and George P. Scriven.[19] Now in his thirties, Mitchell had considered that the time had passed for him to become a flyer. Also, in 1915 Mitchell had severe health problems and was hospitalized for five weeks for both eye strain and what appeared to be rheumatism, an ailment that had affected his father and had bothered him at Fort Leavenworth.[20] When Mitchell returned to duty he found a storm growing in the Aviation Section. Secretary of War Newton Baker had become increasingly irritated with conditions in aviation that affected discipline and the circumvention of regulations. On May 5, 1916, Reber was relieved as chief of the Aviation Section, and it was announced that Lieutenant Colonel George O. Squier, Mitchell's old friend from Fort Leavenworth, would be recalled from his duties in London and made chief of the Aviation Section. This was a critical event for Mitchell because he was named as temporary chief of the section pending Squier's return. It also meant that he now sported the gold leaves of a major. Even the temporary appointment of Mitchell to the position was a matter of concern to airpower enthusiasts.

In the fall of 1916 Mitchell decided to take flying lessons with the object of being rated as a junior military aviator, the first phase of becoming an army pilot. The likelihood that the army would give Mitchell the time to be trained were slim, and there was almost no possibility that his training at a civilian school would be funded. To his credit, Major Mitchell felt that being rated was a necessity if he were to have credibility within a rapidly expanding section. He was able to obtain official orders allowing him to travel to Newport News, Virginia, where he began lessons at the Curtis Aviation School at his own expense. While he could ill afford the cost, Mitchell pressed on with his training.[21] His instructors found him to be very enthusiastic but erratic, capable of first-rate flying one day, poor the next day.[22] He finished his training and applied for junior military aviator status, requesting, at the same time, that the government pay the large sum of $1,470 for the course. The army ruled that government funds could not be disbursed for civilian training for army officers, and Mitchell would have to pay for his own flying school, adding a large expenditure to his financial burdens.

George Squier, who had now returned to Washington, felt strongly that the United States needed to have an observer of aviation developments in Europe. If the United States was heading toward war with Germany, the need was critical. If the Germans decided to resume unrestricted submarine warfare and America found herself at war, there would be a vital requirement that the United States rapidly build an air arm. Squier nominated Mitchell to go to Europe as an aeronautical observer. There were few better choices for Squier in that Mitchell knew Europe, spoke French with facility, and had important connections (through his late father) with men in high positions. Colonel Squier indicated to Mitchell that if things went well, he would not have to wait long before being promoted to lieutenant colonel.

Before Mitchell departed for Paris in late March 1917, he and Caroline decided to close the Washington house and have her return to Rochester with the two children. Of course, William Mitchell was excited with his orders and knew that it would mean a boost to his army career; however, the trip, unspecified as to length, would be expensive, and Mitchell had shown little inclination to economize. It certainly helped that Caroline would be living with her affluent parents for the duration of his assignment.

Mitchell arrived in Paris on March 19, 1917, and began the process of establishing his office.[23] A little over two weeks later the Congress of the United States declared war on Germany, and the United States was now a belligerent, an associate of the Western powers. Suddenly Mitchell became a very important person in Paris, and he opened a large office with space donated by the Paris branch of the American Radiator Company. French citizens and a few American expatriates volunteered to staff Mitchell's expanded operation.[24] With the April 6 declaration of war, Mitchell turned his attention from being an observer of air combat to a participant, and an exponent for expanded American airpower. Mitchell knew, as did any staff officer in Washington, that the nation's first deployment of air in the field had not turned out well. In 1916 President Wilson had authorized a punitive expedition into Mexico to pursue the Mexican revolutionary leader Pancho Villa. The expedition was led by Brigadier General John J. Pershing, who had assigned to his force the 1st Aero Squadron, commanded by Captain Benjamin Foulois. It was found that the United States did not have serviceable aircraft. In April 1917 the United States had only raw manpower and motivation to offer the airwar effort. Mitchell's position as representative of the Air Section of the Signal Corps in Europe took on more meaning than even Mitchell could have envisioned.

Mitchell had always acted with energy and intelligence, and he quickly put all of his abilities in motion. His French contacts were solid, and he called on the Baron Estournelles de Constant, president of the French Senate, for assistance. He then met and cultivated the friendship of the rising politician Pierre Etienne Flandin, who was the undersecretary of state for military aviation. Both men opened doors for Mitchell, and Estournelles de Constant contacted the British air authorities at the British air facility at Vendome, France, to pave the way for a Mitchell visit. He told Mitchell, "If you have the least problem, let me know."[25] Flandin made certain that Mitchell would have the full cooperation of the French Air Service, as he sought out as much information as possible for the American forces that would soon arrive in France.

What kind of air contribution could the United States make if the United States had no real air doctrine? The prevailing view was that air observation was the key to the World War I battlefield, and, given George O. Squier's experience, the United States would lean toward that. Early in Mitchell's dealings with the French and British, he had no preference for airpower because he had no practical combat exposure. Also, much would depend on how General John J. Pershing, the recently named commander of the American Expeditionary Forces (AEF), viewed his air arm.

Given Mitchell's relatively great knowledge about air matters, Pershing would have to rely on him as his authority, and Mitchell would have to establish himself as the most knowledgeable air officer available to Pershing. On April 20 Mitchell began a vitally important tour of the

Western Front. After finding suitable quarters at Chalons, Mitchell reported to the headquarters of the French 4th Army's air section, and in a short time he was at the front, where he was introduced to French air and balloon observation. Mitchell saw very quickly that there was more to air combat operations than just flying against the enemy. He was impressed with French mechanics and with French capabilities in air photography. French air photographic officers showed him what they could do, and Mitchell came away convinced that "concealment from photographic reconnaissance is impossible except in quite heavy woods."[26] On April 23, Mitchell visited a bombardment squadron, where he was told in no uncertain terms that bombardment officers believed, "if they were given enough planes and explosive, there would be nothing left of Germany in a short time."[27] Mitchell was unimpressed with their claims, however, and continued to focus on air observation and the capability of French observation balloons, which appeared to Mitchell to be "dependable and absolutely necessary."

On April 24 Mitchell flew over the lines in the rear seat of a French observation aircraft, thus becoming the first U.S. officer to personally view the battlefield from six thousand feet.[28] A few days later he dined with Maréchal Henri Philippe Pétain, the hero of Verdun, at the French general headquarters. Pétain restated an allied position that France needed manpower and airplanes. He doubted that an American conscript army sent to the Western Front would be of great assistance.[29] Mitchell came away from his hectic tour with a vision for the American air arm. Observation aircraft and balloons occupied his attention, but he stated that "no one can tell me that there is nothing in bombing. It will have a great effect on all operations, if well and efficiently carried out."[30] At this point, he—like many other officers of the time—viewed the application of airpower in terms of support for ground operations. However, unlike most of the AEF staff, Mitchell had been to the front and had actually seen the employment of French airpower, and this made him a valuable asset for Black Jack Pershing.

May was a heavy month for Mitchell. Pershing and his staff, finally scheduled to arrive in Paris in mid-June, would need all of the information they could get, and quickly. The United States Army had gone to war unprepared, and Pershing would have precious little time to get acclimated and begin the process of preparing for troops to land in France and begin training for combat. In May Mitchell was promoted to the rank of lieutenant colonel and designated to be Pershing's chief of air, on a temporary basis, simply because he was already in France and had experience observing French and British air operations. Prior to Pershing's arrival Mitchell also visited General Hugh Trenchard's headquarters several times and heard the British air chief discuss air operations, especially his pet project of a strategic bombing campaign. On June 17 Mitchell again visited General Trenchard's Advanced Headquarters, Royal Flying Corps. There Trenchard restated his theory of strategic bombardment and his concept of massing airpower for operations. He gave Mitchell a copy of a memorandum he had written in September 1916, which he urged Mitchell to read and ponder.[31] Mitchell came away from the meeting still committed to the massing of airpower as a part of ground operations. As far as strategic bombing was concerned, Mitchell believed that it was a good concept to hit and cripple the enemy behind his lines.[32] But he looked at bombing as a way to affect the actual battle, not as a way to destroy civilian morale or to cripple the enemy's war industries. He was still fairly orthodox as far as air warfare was concerned.

Mitchell's stock among the AEF leadership was high during the summer and early fall of 1917. He had a great deal to offer because he had spent time with both the British and the French and had a good grasp of how they worked in organization, in supply, and in combat. Being a gregarious person, Mitchell introduced the staff to prominent Americans and Frenchmen and organized social events for General Pershing and for Brigadier General James Guthrie Harbord, the AEF chief of staff.[33] There were problems for Pershing, however, when he considered what to do with Mitchell. As General Pershing saw it, there had to be a separation of airpower from the Signal Corps, and the chief of the AEF's Air Service would have to carry a general's rank. Mitchell was still young, with practical experience in the field but little experience in the area of administration. It would be a relatively simple matter to promote Mitchell to the rank of colonel, but it would be another matter indeed to give him a general's star. At this point, Mitchell had shown himself amenable to compromise and had manifested an ability to work with others.[34] The stress of combat and the growing commitment to the development of a separate air arm had not yet produced the contentious personality that later marked Mitchell and earned him a number of influential enemies.

Pershing had realized the potential of air reconnaissance during the problematic 1916–1917 expedition into Mexico. He was determined that the AEF develop a viable air arm for service on the Western Front because it was an article of faith with Black Jack Pershing that the United States would put into the field a well-rounded, solidly commanded, modern army that could equal any in Europe. Pershing, however, was also committed to the "cult of the rifle": that the American infantryman with his rifle and bayonet, supported by the sinews of modern war, would be the final arbiter of victory on the battlefield. Mitchell would have to walk a fine line when dealing with Pershing over air matters, and he did so in a meeting with his commander on June 13 in which he discussed his views of tactical and strategic airpower.[35] "Tactical aircraft" would be observation planes, balloons, and pursuit airplanes attached directly to the army; "strategic aircraft" would be a separate force to "attack . . . enemy material of all kinds behind his lines."[36] This marked the evolution of Mitchell's thinking since his arrival in France and after observing the evolution of British air on the Western Front. He did not define how far beyond the forward battle lines the bombardment planes would go.

Mitchell was making solid contributions to the development of the AEF's air arm, and he was living up to his reputation as a dynamic, articulate, hardworking officer. On June 19 he was appointed to the AEF's Aviation Board, which was charged by James Harbord to construct a complete aviation program project for the army in France, to deal with the questions of maintenance and supply, to recommend what schools would be needed, and to formulate plans to place air tactical units into combat.[37] As the second-ranking member of the board and as the officer with the most field experience in France, Mitchell would be instrumental in preparing those recommendations to go to Harbord and then to Pershing. Acting on the recommendations from the board and from officers like Mitchell and Major Towsend Dodd, who was also on the board, Black Jack Pershing began, on his own, to negotiate the construction of aircraft in Europe and to open what would eventually be a vast and expensive air training school at Issoudun, France. The "intrepid airmen" of the Western Front, a favorite of newspaper journalists and popular writers, were in the process of being formed as an image, but they were being forged from nothing, as the United States Army had precious little to offer prior to the summer of 1917. Certainly Mitchell basked in the limelight and the goodwill of Harbord and Pershing even though he still could not wear the wings of an American aviator.

Even so, he continued to fly aircraft in the rear or second seat. With the French maintaining records, Mitchell was in the air almost every day in June 1917, logging twenty-nine hours of flying and making 105 landings. As long as he was in the rear seat with "double controls" he was bending rules only slightly. Usually he flew out of Paris' Le Bourget airfield, and seldom did his flying time exceed fifty minutes, nor did his height go over a thousand feet.[38] In July, however, Mitchell began to fly alone, and on September 23, 1917, he received orders awarding him the status of junior military aviator, backdated to July 19, 1917.[39]

Mitchell had cultivated the friendship of Sergeant First Class Edward "Eddie" Rickenbacker, a daredevil driver who had come to France with

John J. Pershing and his staff on the USS *Baltic*. Mitchell encouraged Rickenbacker to apply for a commission, which he did, and to apply for pilot training at Issoudun. In October 1917 Rickenbacker reported to the massive Air Service training facility and began his journey toward becoming America's "Ace of Aces."[40] Rickenbacker found in Mitchell a kindred soul, an achiever who loved to fly and who relished action. Their friendship would be firm and long-lasting. Whenever William Mitchell left Paris by car on official business, it was certain that Eddie Rickenbacker would be behind the wheel, ready to smash the gas peddle to the floor and try to drive as fast as an airplane could fly.

On June 30 Pershing made Mitchell chief aviation officer of the AEF, and Mitchell had much to deal with. Because the United States had entered the war in an unprepared state, there was a sense of urgency to bring the American army up to European standards. This included air, but severe communication problems between Pershing's AEF in France and the War Department and General Staff in Washington existed because of the natural division between the two. Pershing's aviation board had put in motion plans to open the training center at Issoudun and had coordinated with the British and French regarding training of aviators and ground crews. Washington also had its training concerns and wanted to control the types of aircraft the United States air arm would procure. Making an attempt to coordinate Pershing's AEF and the War Department, George Squier dispatched a mission under Major Reynal C. Bolling to Europe. Mitchell distrusted the intentions of Squier and the War Department in sending the mission because he believed that air questions for the AEF should be settled by the men on the ground who were close to the Western Front, not by those who sat behind safe desks in the War Department in Washington.

The Bolling mission arrived in England on June 26 and stayed until July 2, conferring with British officials on a number of issues affecting the infant American Air Service. British planes and maintenance operations appeared to occupy their time there. When Bolling reported to Pershing in Paris he found that the Aviation Board had already dealt with some of the pressing problems, such as the training of pilots and ground crews. Issoudun would open for classes in the fall, and Mitchell, more inclined to rely on French training methods, resented some of the deals made between Major Bolling and the British. Bolling, in contrast, showed himself to be a competent officer with a pleasing personality, and this Pershing liked. It became evident to Bolling, and he transmitted this to General Pershing, that American production could not possibly begin to supply aircraft until early summer of 1918.[41]

On July 15, Bolling and key members of his mission left France and proceeded to Italy, where the Italians welcomed them with open arms. While in Italy, the mission placed orders for the reliable Caproni bomber, which would be used by the AEF's air arm on the Western Front. Pershing agreed to all of this because, in part, he could control a majority of the air arm's procurement, training, repair parts, and ultimately its deployment at the front. He then decided to retain Bolling and some members of his group in the AEF; they would not return to Washington. Mitchell, still serving as Pershing's staff officer for air, was not especially happy to see Reynal Bolling remain with the AEF. In June Mitchell had suggested to Pershing that the AEF should establish a liaison officer from the AEF to the War Department. According to Mitchell, "This officer should feel himself a part of the army in France. . . . All communications, except routine papers, it is believed, from the Expeditionary Force to the War Department should go through this officer."[42] William Mitchell felt that with Bolling now part of the AEF, the War Department had come to France.

Pershing was faced with a problem in that he now had two fine officers and, since Bolling had been promoted to lieutenant colonel, two air officers of equal rank. He tried to solve it by appointing Bolling to command air activities in the Zone of the Interior, the area behind the actual front. Mitchell was made commander of air activity in the Zone of Advance, the territory that included actual combat operations. On the surface it appeared to be a good compromise given Mitchell's leadership ability and his combative attitude. Bolling, however, was technically competent and seemed to be just the officer to oversee supplies, training, acquisition, and the like. What occurred in reality was a divided authority with no one in charge, and all disputes (and there would be many) were referred to Harbord as chief of staff and eventually to Pershing, as overall commander, if the highly competent Harbord could not resolve them.[43]

On September 3, 1917, Pershing announced that Colonel William L. Kenly, at that time serving as commander of the 7th Field Artillery, would become chief aviation officer of the AEF. Kenly had no aviation experience, but he was a fair organizer and could bring together Bolling's and Mitchell's spheres of influence. To soothe possibly ruffled feathers, Pershing had Mitchell promoted to the rank of colonel even though at that point it was only temporary.[44] Pershing was feeling his way along in creating his modern American army and would find that the selection of Kenly was not an especially satisfactory one, but no American general had ever faced a task of this magnitude. Paris, Pershing found out the hard way, was not the correct place to headquarter the AEF. The distractions were too many, and the costs were too high. He began to look around for suitable headquarters near the front, in a sector where, he hoped, the Americans would soon have their own fighting army. In late August he moved the majority of his staff and AEF functional areas to Chaumont, where the French gave him a very large, but very old, barracks and main building. Mitchell was ordered to move his section to Chaumont, and Raynal Bolling remained in Paris.

Mitchell still harbored some resentment over not being named chief of the AEF's Air Service, but his promotion to colonel before the age of forty and command of the air in the Zone of Advance helped ease his feelings. Like many of the newly arrived airmen of the AEF, Mitchell had a swagger about him. He never let fellow staff officers forget that he had been in Paris to greet John J. Pershing and his novice staff when they arrived. As they moved to Chaumont, where they could hear the distant rumble of guns to the east, every member of Pershing's team wondered what the future held. The average age of the AEF staff officers was forty-three, and they had the faith of youth. It was theirs—and this included Colonel William Mitchell— to win or lose, to do what had not been done before and do it as well as that novice force could.

NOTES

1. Alaska Communications Systems, *49th Anniversary, 1900–1949* (Alaska: ACS, 1949), 3.
2. Mitchell's many efficiency reports are found in the George Hardie Collection in the Golda Meir Memorial Library Archives, University of Wisconsin, Milwaukee, Carton 10, Folder 12.
3. Mitchell to Mother, 6 March 1912, in the William Mitchell Papers, Library of Congress, Washington, DC, Carton 19B. (Hereafter cited as Mitchell Papers.)
4. William Gardiner Bell, *Commanding Generals and Chiefs of Staff, 1775–1983* (Washington: Center of Military History, 1983), 100.
5. Mitchell to Mother, 6 March 1912, Mitchell Papers, Carton 19B.
6. Ibid., Mitchell to Mother, 12 May 1912, Mitchell Papers, Carton 19B.
7. Bell, *Commanding Generals,* 25–28.
8. Mitchell to Mother, 29 October 1913, Mitchell Papers, Carton 19B.
9. Ibid., 25 December 1913, Mitchell Papers, Carton 19B.
10. Ibid., 1 April 1914, Mitchell Papers, Carton 19B.
11. McGovern to Wilson, 24 April 1914, Mitchell Papers, Carton 19B.
12. Mitchell to Mother, 1 September 1914 (possibly written on 8 September), Mitchell Papers, Carton 19B.
13. Ibid.
14. Mitchell to Mother, 21 December 1914, Mitchell Papers, Carton 19B.
15. Report by Squier to the War Department, from London, 26 February 1915, in the George O. Squier Papers, U.S. Army Military History Institute Archives, Carlisle Barracks, PA.
16. Diary entries for 5, 6, 17, and 20 December 1915, in the Squier Diaries,

Squier Papers, officially transmitted to the War Department, U.S. Army Military History Institute Archives, Carlisle Barracks, PA.

17. Squier Report, 26 February 1915, Squier Papers, U.S. Army Military History Institute Archives, Carlisle Barracks, PA.

18. Juliette A. Hennessy, *The United States Army Air Arm, April 1861 to April 1917* (Washington: Office of Air Force History, 1985), 117–120.

19. Papers and Records of the Aero Club of Washington, DC, 1909–1921, Library of Congress, Washington, DC, Carton 1.

20. Alfred F. Hurley, *Billy Mitchell: Crusader for Air Power* (Bloomington: Indiana University Press, 1975), 20.

21. Ibid., 21.

22. Hennessy, *Army Air Arm,* 185–186.

23. Hurley, *Billy Mitchell,* 21.

24. Ibid., 23.

25. Estournelles de Constant to Mitchell, 22 June 1917, Mitchell Papers, Carton 6.

26. Diary Entry, 21 April 1917, in Mitchell's Diaries, Mitchell Papers, Carton 4. One must be very careful in using Mitchell's diaries and memoirs as found in the Mitchell Papers. He tended to revise his manuscripts over the years, but it does appear that his early war diaries, April to August 1917, are essentially original.

27. Ibid., 23 April 1917, Mitchell Papers, Carton 4.

28. Ibid., 24 April 1917, Mitchell Papers, Carton 4.

29. Ibid., 26 April 1917, Mitchell Papers, Carton 4.

30. Ibid., 29 April 1917, Mitchell Papers, Carton 4.

31. Ibid., 17 June 1917, Mitchell Papers, Carton 4.

32. Memorandum by Mitchell to Chief of Staff, 15 June 1917, Mitchell Papers, Carton 34.

33. James G. Harbord, *Leaves from a War Diary* (New York: Dodd, Mead and Co., 1925), 57.

34. Memorandum from Lieutenant Colonel James Palmer to Harbord, 20 June 1917, Mitchell Papers, Carton 34.

35. Diary Entry, 16 June 1917, Mitchell Papers, Carton 4.

36. Memorandum from Mitchell to Harbord, 13 June 1917, Mitchell Papers, Carton 4.

37. Special Orders No. 11, HQ, AEF, Paris, 19 June 1917, Mitchell Papers, Carton 4.

38. An undated document, from French authorities, giving the dates, the length and altitude, the landings, and the destinations of Mitchell's flights. The first flight is dated 31 May 1917, the last 2 May 1918. Mitchell Papers, Carton 40.

39. Special Orders No. 226, War Department, 23 September 1917, Mitchell Papers, Carton 34.

40. Lucien H. Thayer, *America's First Eagles: The Official History of the U.S. Air Service, AEF, 1917–1918* (San Jose, CA: Bender Publishing, 1983), 13, 55, 57.

41. Ibid., 20.

42. Diary Entry, 17 June 1917, Mitchell Papers, Carton 4.

43. Maurer Maurer (ed.), *The U.S. Air Service in World War I,* Vol. I (Washington: The Office of Air Force History, 1978), 54–55.

44. War Department to Mitchell, 29 September 1917, Mitchell Papers, Carton 4.

FOUR

THE WESTERN FRONT

P ERSHING'S DECISION TO MOVE FROM PARIS TO CHAUMONT, the area in which the U.S. Army, when ready, would operate, was a sound one. He took almost all of the staff with him, which guaranteed that Chaumont would be short of space and focus at the beginning. It would take Pershing six months to realize that his staff had to be reorganized and streamlined, with supply and personnel moved to a different location and the combat staff, including operations, intelligence, training, air, and other critical areas, left at Chaumont.[1] William Mitchell understood that if his air command in the Zone of Advance was to function properly, it needed to be away from the crowded main headquarters at Chaumont. On August 18 he requested quarters for himself, twenty officers, one hundred enlisted men, and thirty desks. At the same time he asked for a garage for twenty-five vehicles, an airdrome (airport) for about twenty aircraft, and a location for a meteorological station.[2]

But William Mitchell was not through. He requested that the staff at Chaumont find him "a chateau or large detached house that I could get for my own quarters and for the use of the principal members of my staff."[3] This certainly raised some eyebrows at Chaumont, but Mitchell did obtain use of the Chateau de Chamarandes, which was about a mile outside Chaumont. It was a slightly run-down edifice constructed in the time of Louis XV, with a seedy and neglected garden. The owner was an alcoholic, and his wife had handled the renting out of the building, evidently glad to have a consistent income.[4] At this time Mitchell began using a walking cane, in the British style. Possibly he used the cane because of reoccurring bouts of rheumatism, but many of the Chaumont staff considered it to be an affectation, which was not appreciated.

There was at this time also a growing tension between Mitchell and

William Kenly, newly selected chief of the AEF's Air Service. Kenly had taken a close look at the division of authority between Mitchell and Reynal Bolling and decided that what was needed was centralization of authority in his office. Evidently the staff at Chaumont had convinced Kenly that instruction should be separated from Bolling's control and placed under the staff officer at Chaumont directly responsible for training and instructions. Bolling would return to Paris and be in charge of production and liaison with the allied air ministries. Major Marlborough Churchill, who was then stationed at the chateau with Mitchell, was reassigned as one of Kenly's assistants. William Kenly was aware of the touchy nature of these moves, and he wrote to Mitchell, "As all the men concerned in the future of the Air Service are personally on the best of terms, I am sure that we can all work together until we get the best organized and best running branch of the A.E.F."[5] Mitchell was not that satisfied, however, and contacted Pershing about the matter. In the margin of the letter Kenly sent, Mitchell wrote that Pershing had personally telephoned him to say that all of the changes were in conformance with his wishes.

Kenly tried to be a good Air Service chief, but he was hampered by a lack of knowledge about airpower. Also, there really was not an Air Service, per se, in the AEF; there were lots of hopes and plans, however. With very good reason, the British and the French urged that instruction begin as quickly as possible, prior to the projected late-September opening of the Issoudun training facility. The 1st Aero Squadron (Observation), the "grandfather" of all U.S. aero squadrons, had arrived in France in early September, and the French immediately began their training. For some months, the 1st Aero Squadron would be all that Mitchell had in the Zone of Advance. There was a need for everything to support an air service, including pilots, mechanics, and balloonists, and the time was pressing. Though materiel and facilities were lacking, there was no lack of willing manpower or planning.[6]

In April 1917, when the United States declared war, the nation's army claimed just 131 air officers, and only 56 of them were rated pilots. There was some balloon training at Omaha, Nebraska, but the training of pilots, mechanics, ground crew, and observers was haphazard. The commitment that Pershing made to the British and French in regard to the American war effort dictated that men had to be trained and prepared for combat.[7] Mitchell had very little to do with the day-to-day training of airmen in France. He visited Issoudun, but his primary concern was the development of plans and concepts for actual air combat when the projected hundreds of aero squadrons were ready to fight the Germans.

Having studied the French use of airpower thus far, Mitchell concluded that "the tactics used by the French are defensive in their nature." He, in contrast, argued that the U.S. Air Service should be used offensively. Mitchell believed that the war had taken such a toll on French (and, by extension, British) manpower that it would fall to the Americans to be aggressive and take the offensive, hurling masses of aircraft with fresh, eager pilots at the German air arm, challenging them for mastery of the air over the front. Mitchell restated his belief that the principle of mass, bringing many aircraft to bear in an operation, would assure undisputed success. His focus, as he developed his thoughts, was on the traditional roles of air—observation and pursuit related to the battles along the war's front. Bombing and other potential functions of aircraft were given little consideration.[8]

With characteristic energy, Mitchell also established an air depot to allow training in the Zone of Advance and stockpiling of repair parts. Mitchell and his staff located the depot at Colombey-les-Belles, less than one hundred miles from the trenches of the Western Front. He was also interested in placing airdromes within the American area to support ground combat operations when they began.[9] From July through October 1917, Mitchell was in the air a good deal to observe French operations in the Verdun area and to locate an appropriate site for an airdrome at Amanty, and he flew often to Le Bourget airfield.[10] General Kenly, meanwhile, was in the process of firming up what the AEF could expect from its air arm. The great problem with this approach was that William Mitchell was not a part of the discussions, and he should have been. Kenly, for example, sent Lieutenant Colonel Marlborough Churchill, who had been on Mitchell's staff until Kenly brought him to Chaumont, to discuss air intelligence issues with Colonel—soon to be Brigadier General—Dennis Nolan, Pershing's G-2 (Intelligence).

The first discussions between Churchill and Nolan took place just prior to Kenly's elevation to chief of the Air Service, but Kenly decided to proceed without consulting Mitchell, who in the long run would be responsible for assigning air missions. The basic discussion over what intelligence could be gleaned by air missions was sound. The final document provided for air officers being trained in interpreting aerial photographs to be assigned at army corps level units. Mitchell himself had been impressed with what photographs taken by airplanes could tell the operational planners. It was also fairly obvious that such pictures would assist in bombing missions against critical road junctions, ammunition dumps, rail yards, and troop concentrations.[11] Though it was certainly within the scope of Kenly's authority to assign Air Service officers within the AEF, it was unwise not to bring Mitchell to the discussions. This and other decisions by Kenly soured relations between him and Mitchell and between Nolan and Mitchell. Mitchell came to view Nolan as a meddling enemy at General Headquarters at Chaumont, and subsequent events would prove him correct.

Another area of friction between Mitchell and Kenly came over Mitchell's growing interest in copying British General Hugh Trenchard's concept of an independent air arm in the field. In as early as June, Mitchell had expressed his admiration for the British idea of using air in an independent organization. Trenchard's views, which Mitchell sent to AEF Chief of Staff James G. Harbord, were based on the British experience in air combat from 1914 to 1917. Colonel Mitchell found that Trenchard's emphasis on the attack and on the special nature of air warfare suited his own aggressive personality. The ground would be forced to react to air operations, and the air then would dictate the relationship between the ground forces and the air, especially as American airpower gained control of the skies and was able to mass aircraft against the enemy.[12] Mitchell became more and more convinced that the Air Service, as it was emerging in France, should copy the Trenchard model of an independent arm.[13] Mitchell was moving toward heresy, however, because he was slowly coming to the position that airpower might very well determine the outcome of a battle, or perhaps even a war.

As mentioned earlier, John J. Pershing, who had begun his service with the cavalry, firmly believed in the "cult of the rifle." The combat infantry division as structured by Pershing in 1917 in France was 28,000 men strong, with four large (4,000 men) infantry regiments organized into two brigades. These brigades were supported by machine gun, engineer, quartermaster, medical, and signal units. Artillery was organized into three regiments, and, as the AEF evolved, the division was supported habitually by an aero-observation squadron and a balloon company supporting one observation balloon. In the Pershing orthodoxy, then, air was first dedicated to supplying information to the artillery commander and to the divisional operational planners and intelligence sections. This was a far cry from independence for the Air Service, and for John J. Pershing the key to victory lay in combined arms operations, not independent groups.

Mitchell thus did not endear himself to Pershing's staff at Chaumont. In addition to continually reminding the other officers that he was the "senior" airman in the AEF, having arrived in France even before Pershing, his coolness toward Dennis Nolan, one of Pershing's personally selected General Staff officers, did not help matters. Posing as an "intrepid airman," Mitchell emblazoned his personal aircraft with its own insignia, an eagle in a circle. The officers at the Chateau de Chamarandes saw themselves as a closed shop, open only to the officers of the Air Service. When they could, they frequented local cafés and restaurants together. The pace of development of the Air Service was so rapid that Kenly had a very difficult time overseeing all the components of the Air Service, and, frankly, the priority of action had to be focused on the intensive training going on at Issoudun, in England, and at other facilities throughout France.

By fall, Pershing was anxious that the AEF show some progress in preparation for combat, but by the Christmas season of 1917 there would be only four combat divisions being trained in France. The 1st, 2nd, 26th and 42nd Divisions showed promise, but they were not even close to being committed to battle against a proficient, professional, and tested German army. Furthermore, only a handful of aero squadrons were ready for training close to the front. Pershing's staff was not really functioning well, and there were too many staff sections trying to work in tight quarters at Chaumont. Because Pershing had committed himself to an elaborate air program, the Air Service was a worry to him: "In it there are a lot of good men," he said, "but they are running around in circles."[14] The fact that the Americans were not yet in combat drew criticism in Paris, and when an AEF vehicle ran over a French soldier in Paris some wags remarked that the AEF had finally wounded a soldier, but unfortunately the victim was an allied soldier. Rumors from Chaumont indicated that Black Jack Pershing was ready to change his chief of the Air Service, and Mitchell considered himself to be the natural choice for the position.

It came as a nasty surprise to Mitchell when Pershing instead named Colonel Benjamin D. Foulois as chief of the Air Service. Foulois was a natural choice for the job because of his longstanding work with aviation. Born in Connecticut on December 9, 1879, Foulois had enlisted in the army in 1897, serving in the Spanish-American War in Puerto Rico and then in the Philippines in 1899. In recognition of his outstanding service, Foulois in 1901 was commissioned as a second lieutenant in the infantry, but by 1908 he had graduated from the Signal School at Fort Leavenworth. This brought Foulois into the circle of army officers exploring air questions. He was sent to Fort Myer, Virginia, where he commanded the first balloon the army purchased. On July 30, 1909, Lieutenant Foulois flew with Orville Wright, testing what would become the first aircraft the army would purchase.[15] In 1912 he received his pilot's rating and served in various aviation-related positions until he formed the 1st Aero Squadron and served with John J. Pershing in the 1916–1917 Mexican expedition. Foulois was indeed fortunate, because Pershing became very much aware of what aviation could do in the area of reconnaissance despite the fact that Foulois' thirteen aircraft simply were not suitable for active field service.[16] Pershing had been impressed with what Foulois and his aviators accomplished in the face of severe problems.[17] George O. Squier had relied heavily on Foulois after the United States entered the war in April 1917, and recommended him to Pershing for service in the AEF. Among the considerations that drew Pershing to make Foulois his chief of the Air Service were his record as a team player, his personality, and his acceptance of the Pershing orthodoxy.

William Mitchell deeply resented the appointment of Foulois to the post he felt he should have had. Foulois did not have the experience in the AEF that Mitchell did, but his aviation heritage was longer, his rating as a pilot preceded Mitchell's by five years, and he had held a field command in 1916. Mitchell was determined not to suffer in silence; he went directly to Pershing and complained about being passed over in favor of Foulois.[18] This would now be a difficult situation because Mitchell reacted badly to Foulois, who was promoted to brigadier general. From December on, Mitchell would prove himself to be a tireless worker on one hand, while, on the other, he became less and less of a team player and took every opportunity to snipe at Foulois.

December was not a good month for Mitchell's relationships with the staff at Chaumont. He found another powerful antagonist in Colonel Hugh A. Drum of the staff's training and operations section. Drum had an ego as large as Mitchell's, and his ambition was just as great. Born into a military family in Michigan in 1879, Drum had attended Boston College rather than West Point but was offered a commission in the army (by no less than President William McKinley) after his father was killed in battle in Cuba in 1898. Like so many "Old Army" officers, he saw service in the Philippines, and he served as a staff officer in the 1914 Mexico operation. When Pershing selected officers to sail with him on the Baltic to France, Hugh Drum was one of the favored few. Black Jack Pershing had the highest regard for Drum and marked him for greater responsibilities in the AEF.

Drum already had compiled a number of major assignments when he was posted to the training and operations section in late fall 1917. In December Drum was given the critical task of studying the needs of aviation in France. He came to the task from the perspective of an infantry officer and rejected Mitchell's claims that airpower could very well determine the outcome of the war. "In order to cover their bases, the aviation people are now advocating that we stop bringing the army and apply the tonnage to aviation," he wrote.[19] Drum was convinced that arguments for a massive expansion of the Air Service at the expense of the traditional combat arms were absurd. He became, and would remain, the army's chief spokesman for an air arm subordinate to the overall combat needs of the ground.

In spring 1918, Drum worked with the Operations Section, called the G-3, on a staff study of aviation. The final document bore the stamp of

Hugh Drum and stated his consistent position that "the Air Service is not an independent arm. The employment of the Air Service is, and always must be, based on the combined use of all arms."[20] This was exactly the opposite of the position taken by Mitchell, and it placed the two men at odds. It would not help Mitchell that Drum was close friends with Dennis Nolan and other members of the staff who were selected by Pershing for higher command. Relations were not helped by the severe frustrations over getting Air Service squadrons to the Zone of Advance, and it was irritating to Mitchell that he actually had almost no one to command. In September, for example, the War Department informed Pershing that of all the squadrons being formed in the United States, only eight could be sent to France because there were not enough winter-woolen clothes for the men.[21] In October Pershing requested that eight balloon companies undergoing training in the United States be dispatched to France. The French were ready to train them, and the AEF had purchased Fiat trucks from Italy for their use in the field, but the eight would not arrive until mid-December.[22]

Foulois, who had just been assigned his new position, was as aggravated as any AEF officer over the situation. The large training base at Issoudun was doing the best it could, but too many men were arriving from the United States with few piloting skills. The French and American trainers had to begin the course literally from scratch, rather than take trained flyers and prepare them to function in combat as part of a squadron and survive. A concerned Foulois informed Washington, through Colonel Reynal Bolling, that no more untrained pilots should be sent to France. At a minimum they should have pre-deployment training in gunnery, observation, night flying, and at least some instruction in bombardment.[23] The British had also agreed to train pilots and squadrons, but they also expected the flyers to arrive with a foundation of relevant knowledge. When untrained "pilots" arrived in Britain, there were considerable delays in getting them started in a program because the British felt it necessary to give classroom instruction on the basics of aerodynamics, meteorology, and the like.[24]

In January and February of 1918, Mitchell began to focus on having several aero squadrons start flying missions close to, but not over, the allied lines. The 1st Aero Squadron had completed training with the 1st Infantry Division, and the 12th Aero Squadron, which would become one of the U.S. Air Service's best observation units, arrived at Amanty to begin training with the French.[25] To reach a three-squadron group, the French promised Mitchell and Foulois to add a squadron so that the air arm could have the experience of directing and coordinating a large unit. When activated, the unit would be designated the I Corps Observation Group, and plans were made to have the squadrons flying by spring 1918. The sector they would fly in was known as the Toul Sector, which was basically a quiet area with little ground combat taking place. Because most of the pilots had little or no combat experience, flying in the Toul Sector made good sense.[26] Mitchell was, of course, anxious to get started, and inactivity weighed heavily on him.

Mitchell had, since his student days at Racine College, been a consistent letter writer, but since his arrival in France he focused more on his duties and preparation and less on communication with his wife, mother, and others back in the United States. Caroline remained busy in Rochester with the children, and had, to her husband's surprise, volunteered for the Red Cross Motor Corps in Rochester as part of the war effort. The first phase of her association with the Red Cross began with an intensive course in automotive mechanics. Once she finished that course, she would take instruction in driving an ambulance[27] Earlier in the year Caroline had, on a volunteer basis, taught French to soldiers who were going to France. Mitchell's few letters during this time show that he supported his wife's efforts at home.

Mitchell had much to occupy his mind after New Year's Day 1918. John J. Pershing had resisted French and British efforts to integrate American troops into their own depleted and battle-weary ranks. This process of "amalgamation" had been discussed for months after the United States entered the war, and Pershing had shown himself to be an implacable foe of the concept. He could not do otherwise because his mandate, given to him prior to his departure from American shores in 1917, was to form an American army and to fight as an American entity. Pershing firmly believed in American arms and in the nature of the American character. He would not budge. In January 1918 pressures continued to build on Pershing, and there were those at Chaumont who feared that political considerations in Washington might indeed override Pershing's mandate. Hugh Drum was critical in convincing Pershing to take steps to derail allied attempts to amalgamate.[28] The time had come to form I Corps, AEF, to show that the United States had made progress. As a key member of the operations staff at Chaumont, Drum argued successfully for the formation of the Corps,[29] and to do this I Corps Aviation would have to be formed as well. Mitchell would take command of I Corps Aviation, with orders to prepare to support I Corps ground forces in combat.

Mitchell was pleased to see the 94th and 95th Aero Squadrons arrive for service in the I Corps area. These two squadrons were pursuit aircraft with a current mission to patrol along the lines and to protect airfields and observation aircraft as they trained. Mitchell greeted his protégé Eddie Rickenbacker, who was a member of the 94th. On March 28 Rickenbacker flew his first patrol mission along the lines of the Western Front. Besides having his old friend at the Toul Airfield, Mitchell was concentrating on just what would be expected of I Corps aviation: "The Air Service of an army is one of its offensive arms," he wrote. "Alone it can not bring about a decision. It therefore helps the other arms in their appointed mission."[30] Mitchell had not given up his views on the centralization and the independence of the air, but early in 1918 he remained officially supportive of the dominant view at Chaumont. Two distinct ideas were at the forefront of Mitchell's operational concept: the principle of mass, and immediate counter-air action. The concept was simple: win control of the air.[31] But it followed naturally that if one side had total air supremacy, it could then act with unchallenged impunity against enemy troops on the ground or against enemy lines of supply and communication. In the evolution of William Mitchell's concept it was a short leap from the orthodox view to supremacy of the air arm.

From February through April Mitchell worked on assembling his combat squadrons. A workable Table of Organization and Equipment (TO&E) had been finalized by the end of January 1918, and it included the concept of the air brigade. This was something Mitchell had wanted, since it hinted at a state of quasi-independence for the Air Service at the front. One of those responsible for the TO&E was Major Thomas DeWitt Milling, a close ally of Mitchell who understood the direction in which Mitchell's thoughts were moving. The brigade staff was composed of one general officer, five additional officers, twenty-nine enlisted men, eight vehicles, and three side-car motorcycles. The operations officer was required to be a pilot. It is interesting to note that the brigade headquarters was about the same size as that allocated to the chief of the AEF's Air Service. The brigade commander had at least two wing commanders under his direct command, and these wing commanders each commanded three squadrons and three balloon companies. The pursuit wings, of three squadrons each, came under the control of the brigade commander as well. Wing commanders were required to be rated pilots, and this fit with Mitchell's growing concern that only pilots could command other pilots in tactical units.[32] The organization of the air brigade thus contained distinct echoes of General Hugh Trenchard's earlier concepts.

During those months it was clear that Mitchell could not resist sniping at Foulois, which was particularly irritating to Dennis Nolan and to Hugh Drum. Another person who watched this growing Foulois-Mitchell feud was Major General Hunter Liggett, who in January 1918 took command of I Corps, including Mitchell and his airmen. Liggett had the dubious distinction of being the fattest general in the AEF—an AEF whose commander prized physical appearance. When Liggett and Pershing first met, Black

Jack considered sending Liggett back to the United States because of his poundage; but Liggett, extra weight and all, faced old Black Jack down. He rose quickly in Pershing's estimation and was eventually selected to command the 1st Army, on which Pershing staked so very much. As the I Corps began to take shape Mitchell became chief of Air Service under Liggett, a position that came with the star of a brigadier general. Mitchell was waiting for the promotion, but events would change his prospects considerably.

In March the Air Service became unbalanced by the death of Colonel Bolling. Pershing had used Bolling to take some of the pressure off Foulois and to bring form and structure to the acquisition of aircraft and air-related materiel. His headquarters was in Tours, on the Loire River, and as a result of staff organization Foulois also moved to Tours on February 16, 1918. To keep an Air Service presence at Chaumont, Foulois designated a representative, Lieutenant Colonel Edgar Staley Gorrell, who was no friend of Mitchell.[33] Bolling had done such a good job that Pershing decided to place him in command of II Corps aviation when the corps was formed later in 1918. To prepare Bolling for his transition from the logistical to the tactical, Pershing sent him to observe combat operations. Bolling arrived at the front in the middle of the first great German offensive of 1918, and he found everything in chaos as the Germans made striking advances all along the line. On March 28, Bolling and his chauffeur were fired upon by German troops. Reynal Bolling returned fire, killing a German officer, but was in turn shot and killed. His chauffeur was taken prisoner.[34] With Bolling dead there was confusion at Tours at a very critical time. On March 17 Washington had suspended the shipment of Air Service personnel to France so that combat troops could be sent as rapidly as possible because of the dire emergency along the front. At the same time the training and acquisition of materiel had reached a fever pitch for the AEF.[35]

In late May Mitchell suffered a very personal tragedy when his younger brother, Lieutenant John Lendrum Mitchell Jr., was killed in an air crash at Colombey-les-Belles. John had attended Racine College and in 1913 entered the University of Wisconsin. Upon graduation, with the United States now in the war, he joined the army, was commissioned, and was trained as a pursuit pilot in France. He was assigned to the 95th Aero Squadron, under his brother's overall command. On Monday, May 27, his plane crashed on landing at Colombey-les-Belles, and he was thrown from his aircraft when his seat belt broke. He died almost immediately, and the word of John's death hit William Mitchell quite hard. One observer wrote, "He was a particularly nice boy, much more genial than his brother tho' not so much ability."[36] He was buried the next day at a growing American cemetery north of Toul. When his duties allowed, Billy (as he was now being called by his comrades) Mitchell visited and decorated the grave. To honor the memory of their fallen brother the Mitchell family established a memorial prize in industrial relations at the University of Wisconsin in 1922.

Because Pershing had imposed a strict censorship on the soldiers of the AEF, and the press cooperated with censorship in the name of operational security, newspapers in the United States often wrote "fill" material that bore little relationship to reality. Mitchell's mother received conflicting reports about her son's death in France. The Milwaukee papers carried such headlines as "Former U.S. Senator's Son Dies Fighting the Germans in France," and one story indicated that John Mitchell Jr. was killed by a German bullet or shell.[37] During the first week of June Mitchell wrote a detailed description of his brother's death for his mother:

> He was thrown clear of the plane and never knew what hit him after striking. It was all over. The doctor, ambulance and many of his friends reached the spot almost immediately but it was too late. The damage had been done, his heart was still. The back of his head had been crushed by the impact. Although it is small compensation for us the losses caused the Germans on this day were great.
>
> Aside from being his brother, I have never seen a young man that gave more promise of complete success in anything he attempted. He carried out the best traditions of our family, always wanting to be in the face of the enemy instead of at some station in the interior. . . . I reached the scene of the accident from Toul within an hour of the occurrence. I am writing this on Tuesday morning, with the noise of the motors of the airplanes going to attack the Germans is [sic] almost incessant and artillery fire sounds in the distance. It is hard for me to realize yet that John is gone.[38]

While battle raged, Pershing began to look again at his Air Service. Since Bolling's death there was growing confusion and chaos in the areas of acquisition, training, and coordination with the allies. With so much going on, Foulois was having a very difficult time coordinating everything from the Line of Communication to the Zone of Advance, where Mitchell held sway. Part of the problem rested with General William Kenly and the process of transferring Bolling's authority to Benjamin Foulois. Another part of the difficulty was the delay in orders relieving Kenly, replacing him with Foulois. Kenly simply did not act on issuing orders, nor did he make decisions, because he preferred that his successor have a free hand to establish his own policies. There were, then, three to four weeks when little was done as problems piled up. When Foulois became the chief of the Air Service he did not address the serious undermanning of Bolling's Line of

Communication.[39] Pershing was loathe to simply dismiss Foulois, but he had to have a positive change in the position of chief of the Air Service. He decided to elevate Brigadier General Mason Patrick to the post, which carried a second star. The change would take place on May 29, 1918, and would greatly affect Mitchell.

Pershing was determined to bring order to the Air Service, and he informed Patrick that too many of Foulois's men, many of whom had come from the United States with him, were confused and were working at cross purposes. Patrick found that the U.S. Air Service "was a chaotic condition of affairs."[40] Foulois was to become chief of the Air Service, 1st Army, replacing Mitchell, who was designated to become commander, 1st Air Brigade. Mitchell was furious over being replaced by Foulois, and this change of command aggravated the growing dislike between Billy Mitchell and Benjamin Foulois.[41] Billy Mitchell's attitude and actions had built up a reservoir of ill will at Chaumont, and this now carried over to the change of command at the 1st Army. At that point, however, the 1st United States Army was a name only. Pershing had not received permission to activate the army, nor would he until August 10, 1918, but Black Jack was determined to build the structure that would eventually be an American army.

The acting 1st Army commander, Hunter Liggett, on Tuesday, June 4, ordered that Mitchell was to turn command and all equipment over to Foulois and then leave the area for his new assignment. Major Frank Purdey Lahm, West Point graduate of the class of 1901 and now assigned to the G-3 (Operations) section of 1st Army, observed a rather sorry spectacle at Toul. General Hunter Liggett instructed his chief of staff, Colonel Malin Craig, to carry out his instructions regarding the change of command. What was not needed could go with Mitchell to his new assignment. Lahm recalled that night:

> General Liggett's instructions were definite. Everything was to be turned over that was needed. [Foulois] interpreted it to mean practically everything. Mitchell first named one thing and another—then the personnel, he wanted to keep different ones—in every case [Assistant Chief of Staff, Lieutenant Colonel Stuart] Heintzelman stepped in and repeated the General's instructions. Finally, it came down to his own desk which he said he had had for some time—he was told to keep it, but the men broke it up in trying to move it, so Mitchell said he did not want it.[42]

All of Mitchell's maps and most of his staff were retained by Foulois, and Mitchell then left with a few officers and enlisted chauffeurs and departed for his new headquarters at Neufchateau. Lahm would soon be promoted to lieutenant colonel, and Stuart Heintzelman would become in a few weeks the chief of staff for the newly established IV Corps. This was a bitter moment for Billy Mitchell. His hopes for promotion to brigadier general lay in ruins. Benjamin Foulois, with whom he had been sparring for months, was now his superior officer, and to make matters even worse Hunter Liggett, the army commander, was openly hostile to him. To aggravate the situation even more, Mitchell had been openly critical of Kenly as chief of the Air Service because he was not a flyer. Mason Patrick was the director of construction for the AEF when he became chief of the Air Service. An engineer officer, Patrick had been a classmate of Pershing's at West Point, graduating with him in the class of 1886. Word had reached Patrick about Mitchell's open hostility toward Kenly and Foulois, and Patrick determined that what the air effort in France needed was team players, which Mitchell was certainly not. Hugh Drum had Patrick's ear and was quick to point out Mitchell's arrogance, faults, and often-voiced criticisms.[43] Certainly with enemies at Chaumont, at 1st Army Headquarters, and at Mason Patrick's office at Tours, stories about Mitchell's difficult personality and constant carping would get back to Black Jack Pershing, a man who prized loyalty and teamwork above everything else.

With the United States moving toward the actual activation of the 1st Army for combat, Billy Mitchell was too good an organizer and too aggressive a fighter to be removed from his position as Air Brigade commander. And Mitchell, after twenty years of service, was too good a soldier to shirk his duties when a fight was at hand. In fact a good fight might very well propel him back into the good graces of his commanders, Liggett and Black Jack Pershing. All was not clear sailing for Mitchell once he had been battered by the change of command, and, try as he would, his enemies continued to hit at him. In early July, Colonel—soon to be Brigadier General—Malin Craig, a future chief of staff of the army, tried to tell Mitchell that his "lone wolf" type of command did not fit in with AEF concepts. Craig, then chief of staff of I Corps, sent a stinging memorandum to Billy Mitchell:

> The Corps Commander is convinced that aviation matters in so far as these headquarters are concerned are not being conducted in a manner which will result in the efficiency desired by and expected by all concerned.
> Your own position as Brigade Commander is not construed at these headquarters, under your present orders, as giving you control over the Air units of the corps except as directed by the Corps Commander.[44]

As Air Brigade commander, Mitchell believed that the allocation of aero observation squadrons and balloon companies fell within the purview of his command authority. Colonel Malin Craig obviously thought otherwise, and he demanded that Major Ralph Royce, assigned to I Corps as Air Service commander, direct the observation and balloon units allocation for the corps. In a short and very blunt manner Malin Craig informed Mitchell that Royce would handle all observation and balloon activities in accordance with the direction of the commander and in conjunction with the operational plans of the Corps G-3. The actual duties of the Air Brigade commander and his responsibilities had not been well defined, and Billy Mitchell found himself subject to the corps as well as army commanders. Craig warned Mitchell:

> The Corps Commander desires and insists that personal considerations be obliterated, and that the differences which are constantly arising in connection with Air Service matters cease. To this end, your active cooperation and willingness to meet the views of the Corps Commander are assumed.[45]

Mitchell's relationship with his colleagues had reached a new low. Even so, Mitchell felt that as the Air Brigade commander he had a very large role in the allocation of aero squadrons. He believed strongly in concentration, but ground commanders expected to have an observation squadron and a balloon company assigned to each division and some observation aircraft available to the corps headquarters.[46] One of the critical problems for Mitchell was that he had promised cooperation with the ground forces.[47] The Air Brigade concept, while well-meaning, added a layer of command and control that frankly was not well defined for the air units, or the ground units for that matter. Mitchell's personality and his resentment over failing to get the promotion he felt he had earned by way of experience and seniority complicated the picture as well.

The spring of 1918 was a low point for Mitchell, both personally and careerwise. Still grieving for the recent loss of his younger brother, Mitchell had been greatly disappointed by his failure to become chief of the U.S. Air Service in France. In addition, his lingering financial extravagances still necessitated that he consistently ask Harriet Mitchell for money. Despite the stories of Mitchell being a man of independent means from a wealthy family, his income depended on his army pay and his mother's largesse.[48] From the spring of 1918 on, Billy Mitchell would be a persistent thorn in the side of his contemporaries and his superiors. Mitchell looked forward to future campaigns to win back what he saw as his place in the sun. It never really occurred to Colonel William Mitchell to try to accommodate his colleagues or to admit that men like Foulois or Patrick might indeed be good officers with something to contribute to the growth of the AEF's Air Service.

NOTES

1. See James J. Cooke, *Pershing and His Generals: Command and Staff in the AEF* (Westport, CT: Praeger, 1998), for a full discussion of the shift to Chaumont and the problems in staff organization.

2. Diary Entry, 18 August 1917, in the William Mitchell Papers, Library of Congress, Washington, DC, Carton 4. (Hereafter cited as the Mitchell Papers.)

3. Ibid.

4. Ibid.

5. Kenly to Mitchell, 13 August 1917,Mitchell Papers, Carton 6.

6. See "Report of General William L. Kenly, the Director of Military Aeronautics," *Air Power* (December 1918), 330–335.

7. For an excellent view of training see Sam H. Frank, "Organizing the U.S. Air Service, Part Four, Training Activities in Europe," *Cross and Cockade Journal* VII (Spring 1966), 57–72. One cannot overlook the massive expansion of air training activities in the United States. For a good example see H. D. Kroll, *Kelly Field in the Great War* (San Antonio, TX: San Antonio Printing Co., 1919), and Arthur Sweetser, *The American Air Service* (New York: D. Appleton and Co., 1919).

8. Notes and Memoranda by Mitchell, ca. 29 April 1917, Mitchell Papers, Carton 6. While these thoughts predate Mitchell's appointment as commander of air in the Zone of Advance, his concepts changed little.

9. Lucien H. Thayer, *America's First Eagles: The Official History of the U.S. Air Service, 1917–1918* (San Jose, CA: Bender Publishing, 1983), 27.

10. List of Mitchell's Flights, 1917–1918, maintained by the French, Mitchell Papers, Carton 40.

11. Kenly to Nolan, 27 October 1917, in National Archives, Records Group 120, Records of the AEF, Entry 805.

12. Mitchell to Harbord, 15 June 1917, Mitchell Papers, Carton 34. Mitchell included a memorandum from Trenchard as an annex.

13. Alfred F. Hurley, *Billy Mitchell: Crusader for Air Power* (Bloomington: Indiana University Press, 1975), 35.

14. Mason Patrick, *The United States in the Air* (Garden City, NJ: Doubleday, Doran and Co., 1928), 6.

15. Juliette A. Hennessy, *The United States Army Air Arm, April 1861 to April 1917* (Washington: Office of Air Force History, 1985), 34–37.

16. John J. Pershing, *My Experiences in the World War,* Vol. I (Blue Ridge Summit, PA: Tab Books Reprint, 1989), 159–162.

17. Hennessy, *Army Air Arm,* chapter 9. This chapter contains a vast amount of detailed information about the problems and the successes of aviation in the Mexican operation.

18. Hurley, *Billy Mitchell,* 33.

19. Diary Entry Covering the 1–31 December 1917 Period in the Hugh Drum

Diaries, Hugh Drum Papers, U.S. Army Military History Institute Archives, Carlisle, PA.

20. Memorandum from G-3 Section to Chief of Staff, AEF, 25 April 1918, in the Frank Lahm Papers and Collection, U.S. Air Force Historical Agency Archives, Maxwell AFB, Alabama, Entry 167-601-7.

21. Adjutant General's Office to Pershing, 17 September 1917, in Records Group 18, Records of the U.S. Air Forces, National Archives, Washington, DC, Entry 96.

22. Pershing to the Adjutant General's Office, 5 October 1917, in Records Group 18, Records of the U.S. Air Forces, National Archives, Washington, DC, Entry 96.

23. Bolling to Adjutant General's Office, 23 November 1917, in Records Group 18, Records of the U.S. Air Forces, National Archives, Washington, DC, Entry 96.

24. Typescript manuscript, "A History of the American Air Service in Great Britain, September 2, 1917–December 2, 1918," Mitchell Papers, Carton 34.

25. James J. Sloan, "The 12th Aero Squadron Observation," *American Aviation Historical Journal* (Fall 1964), 179.

26. Sam H. Frank, "Air Service Combat Operations, Part 5, The Tould Sector Operations," *Cross and Cockade Journal* 7, 2 (Summer 1966), 163–165.

27. Biographical Questionnaire, May 1919, Material provided by the Alumnae and Alumni Association of Vassar College, Poughkeepsie, NY.

28. Diary Entry Covering 1–30 January 1918, Drum Papers.

29. Ibid., 30 January–10 February, 1918, Drum Papers.

30. Cited in Mauer Mauer (ed.), *The U.S. Air Service in World War I,* Vol. 2 (Washington: Office of Air Force History, 1979), 175.

31. Thomas H. Greer, "Air Arm Doctrinal Roots, 1917–1918, *Military Affairs* 20, 4 (Winter 1956), 207.

32. Tables of Organization and Equipment, Working Draft, Dated 15 January 1918, from Major Milling's Files. Records Group 120, Entry 805.

33. Mimeographed Study, "Brief History of the Air Service," Dated 1 July 1920, Prepared by the AEF Historical Division, Air Force Historical Agency Archives, Entry 167.401-19, 5.

34. John Howells and Marvin L. Skelton, "Creating General Pershing's War-Time Air Service, Part II," *Over the Front* 14, 2 (Summer 1999), 142.

35. Mimeographed Study, "Brief History," 6.

36. Diary Entry, 27 May 1918, in *The World War I Diary of Col. Frank P. Lahm* (Maxwell AFB: Historical Research Division, 1970), 83.

37. Newspapers clippings citing the stories are found in the George Hardie Collection, in the Golda Meir Memorial Library Archives, the University of Wisconsin, Milwaukee, WI, Carton 11, Scrapbook 3.

38. Mitchell to Mother, ca. First Week in June, 1918, Hardie Collection, Carton 7, Folder 3.

39. Edgar Staley Gorrell to Mason Patrick, 21 June 1919, The Benjamin Foulois Papers, Library of Congress, Washington, DC.

40. Patrick, *The United States in the Air,* 16.

41. Diary Entry, 3 June 1918, Lahm Diary.

42. Diary Entry, 4 June 1918, Lahm Diary.

43. Diary Entry, 24 March 1919, in the Mason Patrick Diaries, U.S. Air Force Academy Library, Colorado Springs, CO.

44. Malin Craig to Mitchell, 1 July 1918, Mitchell Papers, Carton 6.

45. Ibid.

46. For a good summary of what squadrons were indeed available for service at the front see "Battle Participation of the Air Service, AEF, U.S. Squadrons, 1918," in the Air Force Historical Agency Archives, Entry 167.401–19.

47. "Notes on Liaison Between Aircraft and Infantry During Attack," prepared under Mitchell's direction, 28 December 1918, in Mauer, *U.S. Air Service in World War I,* Vol. 2, 205–111.

48. Ruth Mitchell, *My Brother Bill* (New York: Harcourt, Brace, and Co., 1953), 292.

FIVE

VICTORY, 1918

G ENERAL JOHN J. PERSHING HAD CONTINUALLY RESISTED
allied efforts to amalgamate the United States Army into the depleted
French and British divisions. He was stubborn, irritating, intensely committed to his idea that this Great War would show that the Americans, with
preparation and training, could match any army in the world, soldier for
soldier, staff officer for staff officer. As great battles raged all along the
Western Front into July, the allies demanded that the United States send
infantry and machine gun troops to France, and in a hurry. Once fresh
American infantry and machine gunners arrived on French soil, Black Jack
Pershing was pressured to send those units to the field to fight under the
French and British. This Pershing stood against. It was critical that
American divisions have medical, signal, motor, artillery, and other troops
to be a functioning combined-arms combat unit. If there was ever to be a
real American army committed to battle in an American sector under
American commanders, it was vital that Pershing hold the line and build his
force complete division by complete division, despite the shouts and threats
of the hard-pressed and battered allies.[1]

Going into the spring of 1918 Mitchell, commander of the Air Brigade,
could put into the air the I Corps Observation Group, consisting of the ven-
erable 1st Aero Squadron, the 12th Aero Squadron, and the 88th Aero
Squadron.[2] Most of those squadrons had flown with battle-tested French
units until they were experienced and confident enough to engage the
Germans over the front. To command the group Mitchell selected Major
Lewis Brereton, soon to be promoted to lieutenant colonel. Brereton was a
1911 graduate of the U.S. Naval Academy who threw in his lot with the
army's Signal Corps. In 1913 he became an aviator, and in France he had
commanded the 12th Aero Squadron, which would become arguably one of

the best the AEF would put into the air. Mitchell, who had found himself to be rather unwelcome at Chaumont and at 1st Army Headquarters, concentrated on preparing for combat. It was an article of faith for everyone near Pershing that very soon the Americans would be able to conduct their own army-level operational and tactical combat, and Billy Mitchell was determined that American air would play a dominant, if not decisive, role.

The excitement of this possibility pushed everyone harder because the chances of success were great, and the effects of failure were unthinkable. In July Mitchell reconciled himself, for the time being, to being Foulois's subordinate. Mitchell had selected outstanding, albeit mostly combat inexperienced, officers to command and occupy key staff positions, and he tirelessly pushed night flying and night observations.[3] He believed that by developing night flying, a skill the American pilots sadly lacked, success in observation and artillery preparation would be greatly enhanced. Observation that would assist the artillery in target acquisition was a critical task and could not be assigned to daylight flying only.[4] Certainly, concentration on training for night flying and observation to enhance the combat capabilities of the ground units would please those who were Mitchell's most bitter critics.

While Mitchell was doing a good job preparing for the upcoming battles, the perception lingered that he was not a Pershing-styled team player. In many ways Billy Mitchell was a victim of the press as well as a promoter of the romantic view of the "intrepid airman." Pershing had imposed a very strict censorship on the activities of the AEF, and many journalists looking for stories focused on the "gentlemen adventurers" of the air— the pursuit pilots, the aces, and the supposed high life of the aviators. "They were the aristocracy of the late war," wrote one journalist, who told his readers, "They drank plenteously of champagne. The ladies lavished their favors upon them. They slept . . . in comfortable hotel beds."[5] This type of journalism masked the truth. Flyers were in open, bitterly cold cockpits, and many were burned alive when their planes were hit. Many jumped from flaming airplanes to their death from 1,000 feet rather than suffer the horrible agonies of being incinerated alive. Most of them slept in billets near airfields and were often cold and damp in winter. Nights in comfortable hotels under down quilts were few and far between. That was the reality, but few ground troops would have believed it. Unfortunately, Mitchell helped to enhance the picture of the daring, hard-drinking, womanizing pilot. Because so many of the airmen were young, daring, often products of colleges and universities or from nontraditional backgrounds like Eddie Rickenbacker, and were involved in a new and fast technology, they did tend to attract attention and get into trouble.[6] The airfields were behind the trenches, nearer to cafés and to restaurants; there were temptations, and certainly there was visibility. Mason Patrick and Benjamin Foulois had more in common with the soldiers who were enduring training in the trenches, and the stern-faced, hard worked staff at Chaumont had little time for the *beau ideal* view of the aristocrats of the air.

Despite Mitchell's sometimes grating personality and his deteriorating relationship with the Chaumont staff, he remained the best fighting air commander that the AEF had. General Mason Patrick knew that Foulois and Mitchell had a severe clash of personalities, but both men had their roles to play in combat. Patrick was content to leave Foulois to direct training, logistics, and acquisition and to deal with the British and French while Mitchell pushed his pursuit squadrons (27th, 94th, 95th, and 147th) into the air in June to begin to contest the Germans over the front.[7] The pressure on the U.S. Air Service was great, because Pershing was moving as fast as he could to an actual activation of the 1st Army. On May 28, the American 28th Infantry Regiment of the 1st Infantry Division launched a very carefully planned, supported, and orchestrated attack at Cantigny. The attack, which had been planned by Colonel George C. Marshall, went well and enabled Pershing to make a critical point for the British and the French: Americans were ready to fight and contribute more to the battles raging along the Western Front.

Mitchell began to learn a very valuable lesson about air combat, especially about pursuit operations. Many of his pilots, of whom much was expected, had begun to show signs of wear. This situation had occurred before, and without consultation with Foulois, Mitchell decided to rest his pilots. On his own authority he sent three pilots to Paris for a rest leave. Those flyers had shown stress in the air and had crashed their planes on landing due to carelessness blamed on fatigue.[8] At the same time Mitchell clashed with Foulois over the assignment of key officers to various aero squadrons and to staff positions.

According to Mitchell, Colonel Malin Craig informed him that an order had been issued transferring Lewis Brereton from Mitchell's command to the staff of the chief of the Air Service at 1st Army. Mitchell was livid, and he told Craig that the transfer order (which Mitchell had never been shown) was a mistake because Foulois had not discussed the Brereton transfer with him and had no idea of Brereton's impact on the development of I Corps air observation. To remove Brereton now, with all-out combat so near, courted disaster. Foulois was told by Mitchell in no uncertain terms that Brereton could not be sent to the 1st Army because of his work with the Observation Wing.[9] In this particular instance Mitchell was right. It was not good military policy to move a proven commander at that time. Billy

Mitchell was also angered over the lack of transportation for his headquarters and told Foulois that he needed ground transportation for the headquarters staff, for supply, and for couriers. The Mitchell-Foulois feud had become a festering sore, with each man disliking the other with deep intensity.

No one could ever doubt Mitchell's intelligence or his military ability, and he was applying himself with vigor to organizing the air arm for battle. In July he completed an overall scheme to prepare and to structure the Air Service for the fighting that was to come. This organization chart depicted the Air Brigade, the chain of command including a yet-to-be established Air Division commander and staff.[10] How this elaborate scheme would fit into the AEF's chain of command was not stated because, as Mitchell viewed it, the Air Service should be organized as an independent command. An infantry brigade commander in an AEF division carried the single star of a brigadier general, whereas a division commander sported two stars. It stood to reason that as the American army expanded, the Air Service would too, and a brigade structure would soon be outgrown as the American military contribution grew larger. It might be possible that Billy Mitchell could command the equal of a combat infantry division and would have to have an appropriate rank. Pershing had, since his arrival in France, maintained that he should have the authority to assign all officers to higher command with no regard for seniority in the service. In April 1918 he demanded and got the authority to do this.[11] It was then reasonable for Mitchell, and any other officer for that matter, to see the door open for rapid promotion. Those promotions were temporary, through the structure of the National Army, and once the war was over and the National Army disbanded, officers would revert to their original rank. But if one did well as a brigadier or major general during the war, the potential for promotion in the Regular Army would be enhanced.

On July 15 the Germans launched their last great offensive along the Western Front, and American combat divisions temporarily serving with the British and French to gain battle experience were involved in turning back the masses of infantry. A few days after the battle began the Germans were defeated, their offensive powers spent. Counterattacks began, and Maréchal Ferdinand Foch, now supreme commander of allied forces, gave Pershing the green light to activate the 1st U.S. Army. By late August or early September, the United States Army would begin its first independent offensive operation against German positions in the St. Mihiel salient. At the same time, Mitchell's relations with Foulois deteriorated even further. In mid-July Mitchell became irritated again with what he perceived to be continual interference with his command by Foulois and the 1st Army staff.

The cause of this explosion centered on the continual transfer of officers and air squadrons without Mitchell being consulted. The assignment of the 88th Aero Squadron (Observation) to III Corps without his notification deeply angered Mitchell because the 88th had been one of the original I Corps observation groups and had begun flying missions in late May 1918. At the time of the orders, however, the 88th had only six Sopwith aircraft available for duty, no independent photographic section, and no truck to haul radio equipment. As Mitchell pointed out, the 88th had always operated as part of the group, never on its own. To go one step further, Mitchell instructed Foulois that the III Corps "have never had direct control over observation units heretofore [and] they probably will not know what to do with it."[12] Mitchell had a good argument in that he knew more about the readiness of the aero squadrons than did Foulois, simply because he had been working with them for months.

If proper orders had been issued and the chain of command maintained, Mitchell argued, he could have prepared the 88th for the move. He returned to his advocacy of the British system for an independent air arm with its own lines of command and control, and then he lectured Foulois, writing, "I believe it very important that you consider the necessity for regarding the judgement of commanding officers of units on the ground, as distinguished from the judgment on a certain thing formed at a distance."[13] He fumed over visits to air units by Colonel Edgar S. Gorrell, who, Mitchell bluntly told Foulois, did not know anything about the combat situation on the ground or how the air units were configured for combat. Pointedly, Mitchell chided Foulois for allowing officers such as Gorrell to make staff visits without ever discussing issues with the senior commander responsible for those units. He warned Foulois that unless the proper chain of command was observed, the Air Service could experience serious problems as it was tasked for more combat missions. Mitchell was correct in that a staff visit was basically worthless if faults and errors were not immediately discussed and corrected. It did not help any commander if a staff visit resulted in a report that only the staff at Chaumont or at 1st Army read and then filed away.

However, this was hardly the accepted way for a military officer to speak to his superior in rank and authority. The problem had grown out of the extra layer of command and control, the brigade acting between the chief of the Air Service at 1st Army and the units who would do the fighting. The AEF was being built from the ground up by Pershing, and errors would be made as the army tried various forms of command and control. As far as the Air Service was concerned, it was a question of who issued orders, who had actual final authority over units, and to whom subordinate commanders reported. Mitchell's lengthy memorandum to Foulois on July 19 had the tone of a teacher correcting a somewhat obtuse but well-meaning schoolboy, but despite his personal antipathy toward Foulois, he was basically correct. Any unit in combat must be able to react to a clear line of authority or a defined chain of command. Orders must be issued that are clear and coherent; they must state the mission, establish times, and so on. The dangers of two sets of conflicting orders under this system were great. This problem would surface in the infantry divisions as well when the AEF committed to combat and would have at times serious, lethal results.

If the first American offensive were to fail, there was a great possibility that Pershing and many of his generals and colonels would return to the United States in disgrace, and doughboys might very well find themselves in British and French combat units under foreign commanders. Would the American people stand for it? Probably not. Pershing began his planning by directing Colonel Hugh Drum, his 1st Army chief of staff, to assemble a war fighting staff to plan for an offensive against the German-held St. Mihiel salient. Brilliant, energetic, and ruthless, Drum brought the best staff officers to 1st Army.[14] For example, one of his operational (G-3) planners was young Colonel George C. Marshall, who had great experience with the 1st Infantry Division. Unfortunately for the Air Service, Benjamin Foulois was not the AEF's best fighting airman — Billy Mitchell was. Mitchell had organized the fighting air units, had watched his pilots and balloon companies gain experience and confidence, and had seen some of his pilots die, including his own brother. On August 25 Foulois, subordinating his own ambitions to the good of the service and to the upcoming American offensive, requested that Mitchell be made chief of the Air Service, 1st Army because, with the St. Mihiel offensive looming large, the field command needed all of its best fighters.

Hugh Drum did not like Mitchell; but Billy Mitchell was a war fighter and was needed. For Drum, bringing Mitchell to the staff of the 1st Army was akin to inviting a firebug to visit a match factory, and events would prove Drummie, as he was known to his friends, to be correct. The irony of the situation was that Mitchell, like the other commanders, was focused on the battlefield: first gain air supremacy by hard-hitting counter-air fighting, and then mass airpower for a swift, crippling blow aimed at the enemy.[15] When Mitchell arrived at 1st Army headquarters he was ready to prepare an operational plan containing a clear mission statement that would be understood by all, from ground crewmen to group and wing commanders.[16] Despite all his bravado and walking-stick affectations, Mitchell was intellectually prepared to deliver.

The St. Mihiel salient was a triangle-shaped bulge in the lines of the

Western Front. It had existed since the start of the war, and the French had suffered a bloody repulse trying to reduce the salient in 1915. The salient was twenty-five miles wide and fifteen miles deep, with the town of St. Mihiel at the top of the triangle a scant two miles from allied positions. At the base of the triangle was a rail line that was vital for supplying German troops to the south. The large city of Metz, occupied by the Germans since 1914, was also in the base, as were the German-held productive and critical Briey iron mines. Throughout the war the Germans had set up successive defensive positions with barbed wire, machine gun nests, and trenches for infantry. What was not known was that the German high command had made a decision to evacuate the salient, pulling back to near the base of the triangle. With troops in short supply after the bloody March to July offensives, the German General Staff felt it more practical to have a shorter line, which would concentrate their remaining troops to better defend the critical rail line, Metz, and the iron mines. In August, however, Pershing's 1st Army planners fully believed that the Germans would defend every inch of the salient as they had done in the past. There was no good military reason for AEF planners to believe that the Germans would not contest ground they had held for almost four years. Because the salient was shaped like a triangle, American planners saw an opportunity to attack on both sides of the bulge. The AEF would attack in that manner with the best, most experienced divisions they had. The doughboys would be attacking straight into the teeth of prepared defenses manned by battle-hardened German soldiers, or so the 1st Army planners believed. Pershing knew full well when he gave the order to prepare for the attack on August 10 that he was gambling a great deal on the arms and determination of the American soldier.[17]

Mitchell's contribution to the upcoming St. Mihiel offensive would be great, and he began his work with tremendous energy. Wasting no time, Mitchell announced his staff and designated his subordinate commanders, and he could not have picked better. As his chief of staff he had Lieutenant Colonel Thomas Dewitt Milling, a 1909 West Point graduate who had become an aviator in 1911. His teachers were the Wright brothers. In 1913 he commanded the 1st Aero Squadron,[18] and then went on to establish a number of flying records and to participate in working on major improvements in aircraft. Mitchell's operations officer was Lieutenant Colonel Joseph C. Morrow Jr., who graduated from West Point with Milling and had served with him in the 1st Aero Squadron. The pursuit wing commander was Major Bert M. Atkinson, who came to Mitchell's command from the 465th Construction Aero Squadron. His combat experience was obviously very limited, but Mitchell was able to see great potential in Atkinson as a leader of fighting airmen. Major Lewis Brereton took charge as wing commander, Corps Observation Wing, and Major John N. Reynolds, an officer with good command experience, was named as group commander, Army Observation Group. Major Harold Hartney assumed command of the First Pursuit Group, consisting of the 27th, 94th, 95th, and 147th Aero Squadrons. Those squadrons were experienced in combat and were, in August 1918, the best the Air Service had to offer.[19]

With the time for the St. Mihiel offensive growing rapidly near, Billy Mitchell did not mind stepping on toes at 1st Army. Prior to his departure from the 1st Army, Foulois had worked on an assignment list of aero assets to 1st Army and to I, IV, and V Corps. What was produced was basically an equal allocation based on pursuit and observation squadrons and their supporting units. It was a careful but uninspired document, which Hugh Drum published as an order on August 26.[20] Mitchell took a look at the assigned missions on the ground and allocated aircraft based on those missions. If air units were to support an attack, the largest number of aircraft had to be given to the corps making the hardest assault.[21] But what Mitchell envisioned was something that neither Foulois nor Drum had in mind: he was about to bring together, in a massive concentration, enough airpower to deal a knockout blow to the enemy. Pershing himself had opened the door for concentration when, on August 15, he had requested that Foch add seven observation squadrons, nine pursuit squadrons, five day-bombing squadrons, and ten balloons from the French army to give added muscle to the new American air arm.[22] With the addition of French units, and the promise of British units,[23] reaching the much-preached principal of mass was possible.

With Mitchell at 1st Army, there was a new spirit and aggressiveness; there had to be, because the St. Mihiel operation was scheduled to start at dawn on September 12, only two weeks away. Dennis Nolan, Pershing's brilliant intelligence chief (G-2), met with Major John Reynolds, Mitchell's designated group commander, Army Observation. Reynolds had been allocated the 9th Night Observation Squadron, the 24th and 91st Aero Squadrons, a photographic section, and an unnamed French night observation squadron. This group would be at the disposal of Nolan at Chaumont and Colonel Wiley Howell, G-2 at 1st Army. Nolan and Howell were most concerned about the photographic-observation mission that would pinpoint German troops, ammunition dumps, and defensive positions. In a Mitchellesque response to Nolan's and Howell's requests, Reynolds said that three photographic-observation planes would fly in concert, and, given the heavy machine guns on the planes, they "couldn't be stopped in the world war."[24] Nolan certainly hoped that was true, but he warned Howell to be careful in his dealings with Reynolds and with Billy Mitchell. Although his dealings with Reynolds turned out to be very satisfactory, what was not said, and would later become a problem, was how Nolan, the G-2, viewed control and mission allocation for the Army Observation Group, which he saw as strictly a G-2 matter.

While staffs were writing orders and troops from the combat divisions were moving into place for the September 12 operation, Mitchell was assembling what would be the largest concentration of aircraft for one battle during the war. The French had placed at the AEF's disposal a full air division plus several specialized squadrons. General Trenchard had promised Maréchal Foch that the British would cooperate with eight night bombardment squadrons when the tactical situation on the ground allowed. The Italian Air Service would also be present with a few bombing squadrons. Billy Mitchell and his staff were looking at 1,481 aircraft deployed on fourteen major airfields, with 30,000 officers and men ready to do deadly work.[25] To make all of this Franco-American cooperation possible, Mitchell employed Major Paul Armengaud as an assistant chief of staff for Lieutenant Colonel Milling. Armengaud, a French aviation officer, knew how to phrase orders and how to smooth out potential rough spots between the French and the Americans. A decorated member of the French General Staff, Armengaud had been with the Americans since their arrival in France and was personally selected by the French Air Service to be the liaison officer with Mitchell. Pershing had wanted Mitchell to be aggressive when he drew up his operational plans, and that was exactly what Black Jack got.[26]

Billy Mitchell had a reputation of being uninterested in the mundane paperwork of war—writing orders, filing detailed reports, and the like. In preparing for the St. Mihiel operation, however, he brought his considerable talents as a writer to bear on his operational orders, which were clear and concise. On August 20, 1918, Mitchell wrote and sent to 1st Army headquarters his concept of the operation, how he expected the units under his command to perform when the offensive began. He fully understood that Pershing meant for the St. Mihiel campaign to come as a surprise to the enemy, a very experienced enemy with the benefit of almost four years of combat. As Mitchell saw it, it was paramount that allied air first win the critical advantage in reconnaissance by sweeping enemy reconnaissance aircraft from the skies. The Germans would then be blind as to AEF intentions, but the Americans would be able to see the enemy defensive positions and troop dispositions undisturbed from the air. As Mitchell saw it, the overall mission of the Air Service in the weeks leading up to St. Mihiel was to "absolutely prevent access to our lines by enemy reconnaissance aviation." All the while, American aviation had to continue flying over the area of operations, day and night, and take photographs, observe, and report. This fit well with Dennis Nolan's concepts of observation aircraft, and at least Mitchell's concept of the operation initially caused no friction between the two men.

Mitchell divided the Air Service's contribution to the offensive into four separate parts: the preparation, the night proceeding the attack, the first day of the assault, and the exploitation of gains made on that first day. In conjunction with, but not in coordination with, allied artillery, bombardment aircraft would hit strategic targets such as airfields, railroad stations and bridges, and ammunition dumps, and also enemy troop concentrations. This would occur the night before the ground forces went over the top. Mitchell indicated that bombardment aircraft would strike beyond the limits of the artillery, and he counted heavily on the bombers lent to the AEF by the British.

In keeping with Billy Mitchell's combative personality, the greatest tasks for the first day of the offensive fell to pursuit aviation. His fighters were to fly deep into the St. Mihiel salient and attack the enemy's air, breaking up their formations, killing their aircraft, and opening the way for bombardment squadrons to obliterate enemy airfields, destroy supplies, and strike at troops moving to the battlefield. Only when pilots were certain of good air-to-ground liaison with American infantry could they engage in close air support. As Mitchell saw that support, it consisted primarily of attacking enemy positions poised for counterattacks. Getting too close to the actual man-on-man ground combat was dangerous because, in the heat of battle, infantry had the tendency to fire on anything that flew in the air. Because the pursuit aircraft used their machine guns, they had to fly low amid abundant small-arms, mortar, and artillery rounds. It was best, Mitchell wrote, to keep German pursuit aircraft from entering the fight. The best close air support, then, was to sweep German aircraft from the St. Mihiel salient, downing as many of them as the aggressive American and allied pursuit pilots could.

Mitchell firmly believed that the St. Mihiel operation would be a success. It had all of the elements of a well-planned operation: surprise, mass of combat power, well thought-out artillery preparation, use of tanks in the advance, trained and tested combat units with proven staffs and subordinate commanders, and an overwhelming concentration of airpower. In the exploitation phase, Mitchell wrote, the enemy would be in retreat and aircraft would attack enemy ground formations. While all this was going on, the crews remaining on the airfields would move their operations forward to new, selected airfields so that the squadrons could keep up with the advancing infantry.[27] Displacing to new airfields would be tricky because it would involve moving everything from fuel and repair parts to bedding and mess equipment—everything needed to keep the aircraft and their pilots in the air to support the on-going St. Mihiel operation. What an embarrassment it would be if the infantry outran the Air Service!

The Air Service staff planning had to be extremely detailed; it was not enough to count American, British, and French squadrons and give them target areas and missions. This included such concerns as gasoline supplies; availability of spark plugs and other parts; transportation routes into new airfields over what promised to be cluttered, congested roads; bandages and medical assistance, and such mundane but essential items as coffee and bread. All of the plans had to be completed in time to first brief the senior commanders and then give them time to think through the missions assigned. There had to be time enough, too, to plan and present briefings for those who would fly their own aircraft into the teeth of mortal danger. And everyone, from infantry privates to pursuit pilots to generals, was praying for good weather for the battle.

Over a half-million doughboys and 100,000 French soldiers were ready to attack and reduce the salient. The best-trained American divisions were poised; tanks, under the command of Lieutenant Colonel George S. Patton, were in position to support the infantry advance; artillerymen piled rounds to fire; signal corpsmen laid miles of wire; and just after midnight, September 12, a few hours before the artillery preparatory fires were to begin, the sky opened up and rain poured in torrents. From his command post at Ligny-en-Barrois, close to the front, Mitchell exhorted his airmen, saying, "Our Air Service will take the offensive at all points with the object of destroying the enemy air service, attacking his troops on the ground and protecting our own air and ground troops."[28] As the rumble of guns shook the ground, the fighters of the Air Service knew that the work ahead would be serious and deadly. One pilot said of the rain, "This meant just one thing, we would have to perform our duties in dirty flying weather."[29] Just before the skies turned from black to gray, the ground crewmen prepared their craft, and pilots, now soaked and cold, climbed into their cockpits. "We could see the dim ghostly silhouette of the machine as it streaked by, accentuated by the sharp bark and bluish red flames," wrote one squadron commander.[30]

Well-prepared operations orders are based on the commander's concept of how a battle will be fought, reflecting how he views the mission, terrain and weather conditions, the enemy's forces and dispositions, and the condition of his own troops. Billy Mitchell's Battle Order Number One, finished on the afternoon of September 11, 1918, contained all of those elements, and like every fighter Mitchell fully believed that once he committed his forces they would accomplish the missions assigned, driving the Germans from the air and from the field. What neither Mitchell nor his planners could envision was the terrific effect the heavy rains of September 12 would have. After the operation began reports began to arrive at Ligny-en-Barrois that many aircraft were having problems taking off from their airfields. Runways used during the Great War were, in the best of conditions, grassy and level—in the worst of conditions, muddy, rutted strips of ground. As the heavily laden aircraft moved down the muddy runways, many simply sunk in mud or tipped front-end over, their wooden propellers sinking into the morass and breaking. The Air Construction Squadrons had done their best to prepare new airfields for the St. Mihiel offensive, but nothing they accomplished could undo the effects of the heavy downpour. Aircraft had to be in the air if Mitchell's plans were to work. Much rode on the pilots and observers that rainy, chilly day; Billy Mitchell's boasts about airpower and his vision of an independent air arm depended on that mass of aircraft flying over the St. Mihiel battlefield.

American air superiority, combined with the grit and determination of the ground soldier, simply had to win the fight. Before the 94th Aero Squadron (Pursuit) took off to locate the positions of American and enemy troops, its pilots put bundles of cigarettes in the already cramped cockpits, and once over the American infantry they dropped those small packets to the doughboys who were wading through calf-deep mud. Little notes were attached that said "the air service was with them and to keep going like hell."[31] Given the popular view of the airman and his supposed soft billets, availability of French women, and nearness to cafés with an abundance of wine and cognac, it would be hard to believe that the infantrymen were much impressed with the sentiment, but they did enjoy the smokes. Many of the muddy, rain-soaked ground troops saw aircraft downed by enemy ground fire or by German aircraft. What Mitchell, his staff, or the combat infantry divisions could not see was the effect aircraft were having behind German lines. Just the presence of allied bombers or pursuit planes, which had a deadly strafing ability, drove German troops from the roads, slowing the infantry considerably and hindering resupply operations. American after-action reports revealed that often bombs hit the wrong targets, sometimes demolishing civilian homes and the like, but the German soldiers were too battle-tested not to take cover off the roads or away from railroad supply points when the sound of the aircraft was heard. The mass of aircraft during the offensive was having its desired effect.

Other problems arose early on September 12 as the St. Mihiel offensive began. The Day Bombardment Group began the offensive doing little, waiting for specific missions that would allow them to fly deep into the salient and hit targets well behind the front lines. It was not lost on Hugh

Drum and the battle staff at 1st Army that artillery could only fire so far; aircraft extended the killing power of the forces committed to battle. The non-Air Service officers at 1st Army tended to see bombardment in a strictly tactical sense—extending the range of artillery, and affecting the immediate battle. Mitchell, in contrast, by now envisioned a much larger role for bombardment as a deep tactical threat, many miles beyond the armies locked in mortal combat on the ground. The actual mission of bombardment had never really been agreed upon, and some confusion arose when Mitchell's Battle Order Number One defined the mission of hitting enemy divisional and corps command posts in order to disrupt German command and control. Scant attention was paid to a detailed mission statement for the bombers, and Hugh Drum's mission statement only called for "normal bombardment" in the St. Mihiel sector. This was something that would have to be ironed out once victory was achieved over the Germans at St. Mihiel. During the Meuse-Argonne campaign the controversy would surface again, however, as relations between Hugh Drum and Billy Mitchell deteriorated even further.[32]

Another area of frustration that plagued the whole St. Mihiel operation was the coordination between the air observers and the infantry divisions. It was pretty well accepted—although certainly not liked—that infantrymen in the heat of battle would shoot at any airplane and then try to figure out later whether it was allied or Boche. However, before the units went to the trenches, they received instruction on what was then called liaison between ground and air. Each division had an artillery brigade of three regiments, and the balloon and the airplane were invaluable in finding targets of opportunity beyond the human observers' eyes. To coordinate between the ground and air it was vital that there be some method of identifying units, especially the artillery regiments and brigades. Radio communications were primitive, in their infant stages compared to later wars. It was necessary that units on the ground lay out identifying panels, or drop zones, where observation pilots and observers could drop messages. It was then imperative that the artillery on the ground and the aircraft flying overhead have some method of communicating that would be rapid and accurate.

For the artillery the best method of air observation rested with the balloon companies, which had a single balloon with telephone communication with the artillerymen on the ground. Normally a divisional unit had a balloon company assigned to do the important work of target acquisition and fire correction. During the Great War, when "the balloon goes up," accurate and deadly artillery fire followed. The airplane could go beyond the visual observation of the balloon and increase the depth of the artillery fire. One of the great problems of the St. Mihiel offensive was that liaison between airplanes and the ground forces broke down, with some messages being misdropped and others ignored. Billy Mitchell was not especially interested in the balloon with its telephone, and there is no evidence that Mitchell ever went aloft in a balloon, but he and his staff knew better than to overlook its utility. Normally a division had one balloon assigned as a unit in addition to an area observation squadron, and it came as a disappointment that liaison between air and ground during the St. Mihiel offensive was not better. After the conclusion of the St. Mihiel operation Major Lewis Brereton wrote a scathing report to Billy Mitchell that blasted both the air observers and the corps and divisional artillery units.[33] Mitchell had achieved his desired mass, which was spectacular and did great service, but there were failures in the observers-divisional artillery cooperation that time would not allow the AEF to correct.

St. Mihiel was Mitchell's first great professional victory, that one moment that could make a man's career worthwhile. The attack came as a surprise for the Germans, and on September 12 and 13 the Americans had almost total air supremacy. On September 14 the Germans recovered and sent more pursuit aircraft to contest the Americans, but still the allies outnumbered and outgunned the enemy, who had only 300 aircraft available and were unwilling to dedicate more to an area they had decided to withdraw from. Mitchell had achieved mass and used it well. His summary of operations for September 12 and 13 reveals an aggressive attitude by all of his squadrons. Reconnaissance and infantry contact flights dominated the activities, in keeping with Mitchell's belief in support for the ground attack.[34] In every battle order published by Mitchell's headquarters at Ligny-en-Barrois there was a restatement of the vitally important use of air support for the troops on the ground. Billy Mitchell was at his leadership best. Despite the miserable rain and cold conditions in the open cockpit, he was in the air many times to observe his squadrons moving into combat. The expert use of German artillery and placement of machine gun positions inflicted casualties on the allies, but it was not enough to halt the doughboys and French in their attack to reduce the salient. The operation was well supported by air reconnaissance and contact patrols; bombing of road junctions, ammunition dumps, and rail stations; and strafing of German positions.[35]

The stress on the staff of the 1st Army was understandably great. Many of the key players in the St. Mihiel offensive got little sleep as reports continued to pour in from the units in contact with the enemy. Tempers were short and nerves were frayed, and the atmosphere at headquarters was electric because Black Jack Pershing had staked so much on this operation. While combat divisions drove toward their objectives in the St. Mihiel salient, other troops were on the move toward staging areas from which they would begin the next fight, in the Argonne forest.

The same air of tension existed at Ligny-en-Barrois, where Mitchell and his staff monitored reports from the aero squadrons fighting above the salient. There were concerns over supplies of gasoline, oils, and repair parts, but the detailed Airplane Gas and Oil Report filed at the end of the fighting indicates that, at least for this operation, fuel and oils were never in short supply.[36] The construction squadrons of the Air Service were moving on to prepare new airfields to support the upcoming Meuse-Argonne campaign, and like their comrades in the ground forces they were part of the gigantic and confused traffic jams that clogged the already overtaxed and not very well maintained road system. Mitchell remained at Ligny-en-Barrois during the duration of the St. Mihiel operation and did not travel to 1st Army headquarters, where his presence would not have been appreciated by Hugh Drum and other staff officers. Pershing was aware of the progress of the air war over the salient and the light losses sustained by Mitchell's airmen, and he had every reason to be pleased by the reports from every component of the 1st Army. These very satisfying reports, however, were to cloud Black Jack's vision of the next fight, as we shall see.

Mitchell's own after action report, sent to headquarters on September 21, 1918, was free of bombast and the usual Mitchellesque rhetoric. In a straightforward description of the operation he discussed the weather conditions, the types of missions flown, and some of the problems incurred in maintaining contact between planes in the air and the antiaircraft and combat forces on the ground, including a few of Brereton's cogent remarks about the failure of air-ground liaison during the operation. He pointed out that pursuit tactics could very well advance to strafing enemy positions and hitting the Germans with small antipersonnel bombs. He had observed the psychological effect that enemy aircraft had on American troops and, from reports from his own pursuit pilots, he knew that they played havoc with exposed combat-experienced German forces as well. The inexperience of American bombardment crews, Mitchell wrote, caused problems when they were sent on missions deep into the salient at high altitudes. What Mitchell then suggested was that, as the bombers learned their trade, they could be better employed closer to the front lines where their bombs could have the best effect against the enemy. All of the plans, all of the preparation, and all of the daring and courage of the airmen brought about the desired results: air superiority over the main battle area, which was, in Mitchell's mind, one of the keys to victory in the recent operation.[37]

Pershing was so pleased with the overall operation that he recommended Billy Mitchell for promotion to the rank of brigadier general. He also decided to recommend Mitchell for the Distinguished Service Cross (DSC) for valor. The narrative for Mitchell's DSC cited heroic action at Noyon on March 26, 1918, during operations along the Marne River in July 1918, and for his conduct above and beyond the duties of an army Air Service chief during the St. Mihiel operation. From September 12 to 16, Mitchell had flown his aircraft daily over the lines, sighting German troops and positions, which he reported directly to the American infantry units attacking against them.[38] For Pershing this was proof positive that the Air Service was best used when in support of the ground troops in accomplishing their missions. Billy Mitchell's stock was never higher, and Pershing sent Mitchell a personal letter of congratulations "on the successful and very important part taken by the air forces under your command in the first offensive of the 1st American Army. [It] . . . is as fine a tribute to you personally as the courage and nerve shown by your officers [is] signal proof of the high morale which permeates the service under your command."[39]

When the St. Mihiel campaign came to an end on September 16 the AEF had to immediately turn its full attention to the Argonne forest, but time was short. Pershing had promised Maréchal Foch that the Americans could fight at St. Mihiel and, at the same time, prepare for a major battle in the Argonne, scheduled to begin on September 26. The task before Pershing and his planners was staggering, to say the least. He had committed his best divisions—1st, 2nd, 5th, 26th, 42nd, 82nd, 89th, and 90th—to Hunter Liggett's 1st Army at St. Mihiel, and they did their job beyond all expectations. But the 1st Army was Pershing's first team, and the units selected to begin the Meuse-Argonne operation were not as well trained and had seen less time in France. In fact, some of the infantry were poorly trained for battle.[40] The AEF would pay a serious price for the promise to end one fight and be committed to another battle in less than two weeks.[41] The euphoria over the St. Mihiel victory at AEF headquarters at Chaumont obscured the serious weaknesses in planning for the fight for the Argonne.

Mitchell also had to adjust, and adjust quickly, to prepare for the upcoming campaign. He would have about eight hundred aircraft available for combat, and two hundred of those were French airplanes.[42] During the St. Mihiel campaign 1st Army Air had only twenty-one observation balloons available for service, six of them French. A number of lessons were learned about the use of balloons as well as aircraft. The balloons, despite the weather and enemy air attacks, accomplished their two stated missions: reconnaissance of a sector, and adjustment of artillery fire.[43] The U.S. Air Service emerged from St. Mihiel tactically sound, and Mitchell preferred to leave his command structure intact for the Meuse-Argonne offensive. In planning for the upcoming operation Mitchell could count 845 airplanes sitting at various airfields, but only 670 were fit for immediate combat. The remainder needed repairs and maintenance.[44] Pressure now fell on the supply system for the 1st Army air units, and this was complicated by special needs for aircraft. To facilitate the flow of critical supplies, especially gasoline, oils, and repair parts, Mitchell and Milling decided to place the supply base for the Argonne operation at the 1st Air Depot at the Colombey-les-Belles airfield. While the fighting was in progress from the Meuse River through the Argonne forest there was a need to anticipate possible operations or a possible German counterattack in or near the old St. Mihiel battlefield. Consequently, another base was established at the Advanced Air Depot at Behonne, west of St. Mihiel.[45] The great concern was the flow of spare parts, gasoline, and oil for the Air Service; if the distribution of these critical supplies slowed down due to bad roads or road congestion, the Air Service might not be able to fly missions.

Mitchell was too good a commander not to be concerned about supply, but he felt more comfortable preparing plans and leading his airmen by example. Whether its missions were observation, pursuit, or bombardment, a group consumed 1,500 gallons of gas and 400 gallons of oil per day. New airfields had to be constructed, as well as billets for the troops, mess halls, and repair hangers. All shifting of air units had to be done at night to mask the moves from German air observation. The counter-reconnaissance battle was begun again to blind the Germans to American preparations for the massive attack into the Argonne. Repositioning was a monumental task, and it had to be completed prior to the start of the offensive, scheduled to begin on September 26. Mitchell moved his command post to the airfield at Souilly, which was less than twenty miles from the ground forces' line of departure. He pushed the bulk of his force to the west of the old St. Mihiel area because his aircraft could carry fuel for a little over two hours of flight. To accomplish the missions assigned to the Air Service it was vital to have the new airfields as close as possible to where the infantry and the artillery would fight. As the Americans advanced—rapidly, it was hoped—Mitchell would have to displace airfields to support the ground attack.[46]

While his aero squadrons were moving to the west, Mitchell cemented his relationship with the Air Officers of the 2nd French Army, French 2nd Colonial Corps, and French 17th Army Corps. French Army-level air commander Major de Vergnette would also command the U.S. Air Service units attached to the U.S. IV Corps. It was agreed that the French would operate under orders issued by Mitchell. This was a critical arrangement in that it gave to Mitchell's force those all-important French squadrons with 200 planes, and the French would rely on their own supply system for fuel, repair parts, and maintenance.[47] Once that was secured, Mitchell turned his attention to his own battle orders for the Argonne operation. His guidance was the same as it had been for St. Mihiel: "Our Air Service will take the offensive at all points at Daylight September 26, 1918, with the object of destroying the enemy's air service, attacking his troops on the ground and protecting our own air and ground troops." His subordinate commanders— Reynolds, Hartney, Atkinson, and Pagelow (balloons)—remained the same. There was no reason to change perfectly effective horses in the middle of a stream.

The problems with the Argonne fight, and there were many, began with the euphoria over the victory at St. Mihiel. Pershing had committed his trained and tested combat divisions to the fight because it was imperative that the AEF win that battle. While the Grand Old Men of the AEF were slugging it out over muddy terrain against an enemy who had decided to abandon the salient, American divisions were moving into staging areas for the next fight. Incredibly, just days before the opening of the campaign, a number of those novice divisions saw their most promising staff officers assigned to attend the AEF's General Staff College, leaving critical staff functions to subordinates who had little or in some cases no experience in divisional level combat operations. Pershing himself had just finished the St. Mihiel fight, and he and his staff had almost no time to reorient their thinking and focus on the new terrain over which they would fight. This notwithstanding, Pershing had a mass of willing men with high morale and an ignorance of the conditions that awaited them in the Argonne forest.[48]

The Germans had decided to defend every inch of ground in the Argonne, and they had well-prepared defensive lines with excellent observation and first-rate troops. At St. Mihiel the German commanders fought with well-placed machine gun positions and well-served artillery; they did this to cut their losses in a fight over an area they had decided to abandon. But the enemy high command believed Argonne could extract a terrible price for terrain gained by the Americans. Students of military history know that an attacking force has to have three or more attackers to every one defender, and the German commanders were well aware that the odds favored a well-designed defensive force supported by first-rate artillery. The entire campaign would be fought in rainy conditions with the temperature continually dropping, and as the bitter struggle continued the specter of a winter in Eastern France loomed large for the AEF. Unbelievably, some American divisions moved into their attack positions still in their light canvas summer-weight uniforms. Losses to disease and the elements would become alarming. Even so, Pershing was ready to hurl the untested portion of the AEF against the tested German troops in the Argonne.

The attack opened in rainy weather on September 26, and it seemed at the start that the Argonne offensive would go as well as the St. Mihiel attack. Enemy aircraft, however, were much more active over the battlefield than they had been during St. Mihiel, since the Germans had decided to defend the area to the last. They had good troops in large numbers and well-prepared defensive positions with overlapping fields of fire for their machine guns, which were well-served by determined artillery. The terrain over which the AEF would attack was an infantryman's worst nightmare, with high ground belonging to the enemy and with heavily forested areas, deep ravines, and poor roads. The attack slowed as many of the untested AEF divisions found that they were not prepared for sustained combat. The old, battle-tested divisions that had fought in the St. Mihiel offensive were either still on the road to the Meuse-Argonne or were refitting on the ground they had conquered during the September 12–16 fight. Pershing had to slow the fight down and reorganize his forces to regain the momentum of the assault. Tempers were short at Chaumont and at 1st Army headquarters as Pershing set a target date of October 4 to renew the attack.

The intelligence sections at Chaumont under Dennis Nolan and at 1st Army under Wiley Howell wanted more reconnaissance and photographic missions flown to support the renewal of the offensive. Antagonism had been building for some time between Mitchell and Hugh Drum and Dennis Nolan. Drum and Nolan did not like Mitchell's swagger, nor did they appreciate his emphasis on the emerging Air Service as a major force on the battlefield.[49] Wiley Howell continually requested, as he should, air reconnaissance in preparation for the renewed attack, and often he was informed by Mitchell that poor weather simply prohibited such missions. Howell went to Nolan, who exploded and demanded a showdown with Mitchell and Drum. One of the problems for the three highly aggravated officers was a question of doctrine, or who actually set the missions for air reconnaissance. Nolan firmly believed that it was the province of the G-2 to task the Air Service and order missions. Mitchell, in contrast, argued that the 1st Army Air Service commander, knowing the condition of his men and aircraft and under what conditions they could or could not fly, should have final say over missions. During a testy meeting of the three men, Nolan observed that "Mitchell had one excuse after another; he didn't want to do it; he didn't want to do the routine photographing and observation missions that are absolutely essential to the workings of intelligence." Drum seized the moment and instructed Mitchell to cooperate with Nolan and Howell, but "Mitchell was straightened out on it and he laughed it off as usual."[50]

The antagonism between Mitchell and the 1st Army staff was heightened by the fact that they were all engaged in a great world war, a war so vast that it eclipsed the great American Civil War in manpower, technology, and lethality. Soldiers train for war, and since 1865 no officer had had the opportunity to command troops in great battles. The size of the AEF infantry divisions of 28,000 officers and enlisted men matched the strength of an army corps in 1865, and those divisions contained everything from infantry and artillery regiments to motor transport, sophisticated medical units, engineers, and signal units that could link the commander to his forward-fighting units by telephone. A commander could instantly learn of the progress of the battle; no longer would field generals have to dash on horseback into the shot, shell, and smoke of the battlefield. The airplane and the balloon added a third dimension, that of height, to the field, and a commander now could see further than any previous fighter. It would thus come as no surprise that ambitious, competent officers such as Billy Mitchell and Hugh A. Drum wanted to be a part of this conflict as a commander of soldiers.

Billy Mitchell had a command, a prize, the position of chief of the Air Service of 1st Army, and a record of great achievement in the St. Mihiel fight. Drum was the consummate staff officer, the detailed and meticulous planner, without whom no army could survive for very long. Dennis Nolan got his chance to win his spurs when on September 28 he was sent to command the 55th Infantry Brigade of the 28th Infantry Division, a Pennsylvania national guard unit. During the heavy fighting around a small town named Appremont, on October 1, Nolan earned the Distinguished Service Cross for valor under fire. Drum had spent a good deal of time with Major General Charles T. Menoher's hard-fighting 42nd Division, the Rainbows. Drummie had a real sense of what the doughboy had to endure in the trenches, but that was a far cry from actually making those life-and-death decisions while in command of soldiers. Without a field command there would be little chance of being known as a battle warrior, and there would be no Distinguished Service Cross for valor under hostile fire. In peacetime training and even more in combat, there are strains between staff officers and field officers. While their ultimate goals are the same and while one could not exist without the other, the strains are there. The staff officer works the current battle and looks beyond to the next, and he solves problems in logistics, personnel, transportation, and the like. He is the one who, with his commander's intent as guidance, tasks the combat units with missions.

On the other side of the coin, the field officer deals with the immediate battle, deals with the loss of his men, and cannot look beyond the lethal fight raging before him. These conditions, coupled with personality and ambitions, fueled the fire of dislike between Billy Mitchell and Hugh Drum. Mitchell saw himself, with good reason, not as a staff officer but as a commander of men, and many of those men, including John L. Mitchell Jr., did not survive the conflict. Billy Mitchell was continually in the air sharing, albeit from a distance, the dangers of flight and air combat. When confronting Hugh Drum, his greatest enemy (and he would remain so), Mitchell had the field commander's attitude that he knew best because he had been there and Drum had not. Clashes between the two soldiers were great and would have long-lasting consequences, but perhaps they were unavoidable.

While Mitchell's flyers continued to fly reconnaissance and protection missions, plans were made to repeat, on a smaller scale, the concentration of airpower he had achieved in a spectacular manner during the St. Mihiel operation. Mitchell was able to assemble two hundred bombers, with one hundred pursuit planes for protection, and prepared a bombing campaign to disrupt German concentrations of troops in the Meuse-Argonne area of operations.[51] On October 9 almost eighty-one tons of bombs were dropped on the enemy at the cost of one allied plane.[52] Mitchell was ecstatic over the results. He later recalled, "It was indeed the dawn of the day when great air forces will be capable of definitely effecting a ground decision on the field of battle."[53] With so many planes in the air, Mitchell went to headquarters and asked Drum and Nolan to come outside to see them flying toward their targets.[54] Nolan and Drum were unimpressed with the display of airpower, though, and Hugh Drum later disputed Mitchell's claims that his aircraft broke up German troop concentrations on the ground, saying that "nothing of this sort happened."[55]

Regardless of how Nolan, Howell, and Drum felt, Mitchell's contribution to the war effort merited him promotion on October 13 to the rank of brigadier general. On the same orders Hugh Drum was also appointed to this rank.[56] Mitchell's swagger, his obvious pride in his airmen, and his continued emphasis on the development of an independent air arm had aggravated high-ranking staff officers, but he continued to command with daring and concern for his men. Because of the stress of flying in bad weather and with increasing German opposition, Mitchell directed his subordinate commanders to evaluate the mental and physical condition of the pilots and observers. If they showed signs of stress they were to be sent on leave for a few days.[57] There were loud complaints from staff officers at Chaumont and at 1st Army headquarters about the so-called coddling of flyers. Why should not a lieutenant or captain in an infantry company or an artillery battery fighting in the Meuse-Argonne be sent away for a rest? Major General Mason Patrick, as chief of the AEF's Air Service, paved the way for Mitchell's order in August when he announced that Countess Fenelon had offered her chateau as a rest area for officers of the Air

Service. Patrick's guidance was simple. If a flying officer showed signs of stress from combat operations and a medical officer concurred, he could be sent to the chateau, which was thirty-five kilometers northeast of Chaumont. There was an Air Service rest chateau near Versailles as well.[58] Critics of Mitchell's rest policy failed to point a finger at Patrick, who, by the way, never saw fit to question Mitchell's actions.

The Meuse-Argonne offensive had become a bloody slugfest between the AEF and determined German defenders. Many good divisions, such as the 1st, 42nd, 77th, and 82nd, were used up in a few days of combat. Pershing's belief in the "cult of the rifle" meant that casualty lists were long, and some aero squadrons experienced heavy losses in pilots, observers, and aircraft. Aircraft losses were critical because the flow of machines from the United States had dwindled. In late August Pershing had reached a high level of frustration with the output of American airplanes. He wrote to General Kenly in Washington, "It is apparent that we can not count upon your being able to send any single seaters or biplane pursuit planes for some months to come and we must rely upon our allies to supply such planes for present."[59] But allied air production also had to serve its own forces, and replacement of lost aircraft became a serious problem. By mid-October it was clear that the AEF might soon reach the limit of its contribution in manpower and in aircraft for 1918; air combat, weather conditions, and accidents had all taken their toll on the fragile airplanes.

At that time Pershing made a far-reaching decision. He gave up command of the 1st Army and created the 2nd Army to simplify the command structure. Being head of the AEF and commander of the 1st Army was simply too much for one officer to handle. Now Pershing was General of the Armies, and he made Mitchell chief of the Air Service for an army group, a new and extraordinary position. Colonel Thomas D. Milling took command of air for the 1st Army, and Pershing selected Colonel Frank P. Lahm to command the air in Robert Lee Bullard's 2nd Army. Lahm had commanded the Balloon School at Fort Omaha, Nebraska, in 1917. He arrived in France with Reynal Bolling, and Pershing refused to send him back to the United States.[60] Colonel Lahm was a team player as well as a competent air commander, and the staff at AEF headquarters liked him and relied on him to solve air problems. Lahm did not like Mitchell, nor did Mitchell like him, and they had clashed several times in the proceeding months. Lahm believed that Mitchell showed little interest in the Air Service of the 2nd Army because of this mutual dislike.[61]

As air chief for an army group, Mitchell remained fairly orthodox in his approach to the battlefield. In his first published memorandum to the armies, Mitchell stated, "The duties of the Chiefs of the Air Service of the armies are to see that the aviation assigned to armies works in the closest possible liaison with the troops on the ground. Every effort will be made to make this a success, particularly with the staffs, infantry and artillery."[62] The AEF had only a few days left in combat when Mitchell gave his guidance on October 26. German forces were finally tired out and were in retreat all along the line. Seventeen days later, at 11 A.M., November 11, 1918, the guns of the Western Front fell silent as an armistice went into effect. The war was not over—the final peace had yet to be negotiated and signed, and Pershing had agreed to send an occupation force to the left bank of the Rhine River. There were, of course, congratulations all around, and Mitchell visited Milling's and Lahm's headquarters and celebrated with the airmen who had contributed so much to the victory, if the peace held. November 11 did not come too soon for the AEF or the AEF's Air Service, because planes were in dire need of intensive maintenance and pilots and observers were worn out.

World War I had been a major turning point in Billy Mitchell's army career. He had arrived in France as a lieutenant colonel prior to the declaration of war, and eighteen months later he was a brigadier general and chief of the Air Service, Army Group. Mitchell was the most experienced fighting airman in the United States Army. He had earned a Distinguished Service Cross for valor in the air and had established himself as a commander who would lead his men by example. Certainly Billy Mitchell was no desk officer, and the men under his command respected, if not idolized, him. He also had a number of serious enemies, including Hugh Drum, Dennis Nolan, and Malin Craig. Pershing recognized that Mitchell in war had few equals, but Pershing was a general who prized loyalty and team-playing, and as far as Black Jack was concerned the jury was out on Billy Mitchell. Could Mitchell return to the United States and function in a peacetime army? Certainly his energy, his intelligence, and his motivation were exemplary, but could they translate into a functioning staff officer rather than the dashing, intrepid airman of the Western Front?

NOTES

1. John J. Pershing, *My Experiences in the World War,* Vol. 2 (Blue Ridge Summit, PA: Tab Books Reprint, 1986), 10–14.
2. Sam H. Frank, "Air Service Combat Operations, Part 5, The Toul Sector Operations," *Cross and Cockade Journal* 7, 2 (Summer 1966), 163–164.
3. Mitchell to Foulois, 9 July 1918, in Records Group 120, Records of the AEF, National Archives, Washington, DC, Entry 805. (Hereafter cited as RG 120.)
4. For a good survey of the whole question of night flying during the Great

War see William Edward Fischer Jr., *The Development of Military Night Aviation* (Maxwell AFB: Air University Press, 1988).

5. Clinton W. Gilbert, "Mitchell and the Air Service," *Review of Reviews* 71 (April 1925), 375.

6. Mauer Mauer (ed.), *The U.S. Air Service in World War I*, Vol. 1 (Washington: Office of Air Force History, 1978), 89–90.

7. Mason Patrick, *The United States in the Air* (Garden City, NJ: Doubleday, Doran and Co., 1928), 24–25.

8. Mitchell to Foulois, 12 July 1918, RG 120, Entry 805.

9. Ibid.

10. "Organization of the Air Service, Zone of Advance," ca. June-July 1918, in RG 120, Entry 632 (G-2, Air).

11. Pershing to General Peyton Conway March (Chief of Staff of the Army), 12 April 1918, in Records Group 18, Records of the U.S. Air Forces, National Archives, Washington, DC, Entry 96.

12. Mitchell to Foulois, 19 July 1918, in the William Mitchell Papers, Library of Congress, Washington, DC, Carton 6. (Hereafter cited as the Mitchell Papers.)

13. Ibid.

14. See James J. Cooke, *Pershing and His Generals: Command and Staff in the AEF* (Westport, CT: Praeger Publishing, 1997), 100–101, 127–128, 130.

15. Thomas H. Greer, "Air Arm Doctrinal Roots, 1917–1918," *Military Affairs* 20, 4 (Winter 1966), 207–208.

16. William Mitchell, "The Air Service at St. Mihiel," *World's Work* (August 1919), 361–366.

17. Pershing, *My Experiences,* 225–226.

18. For more on the 1st Aero Squadron see Mauer Mauer (ed.), "The First Aero Squadron — 1913–1917," *Aerospace Historian* (October 1957).

19. General Orders No.1, 1st Army Air Service, 28 August 1918, RG 120, Entry 805.

20. Special Orders No. 78, Headquarters, 1st Army, 26 August 1918, RG 120, Entry 805.

21. Stationing List, 1st Army Air Service, 15 September 1918, RG 120, Entry 805.

22. Pershing to Foch, 15 August 1918, in the Hugh Drum Papers, U.S. Army Military History Institute Archives, Carlisle Barracks, PA. (Hereafter cited as MHI, with appropriate collection cited.)

23. Foch to Pershing, 27 August 1918, Hugh Drum Papers, MHI.

24. General Nolan's dictation for an unpublished book, given between 15 October 1935 and 31 January 1936, in the Dennis Nolan Papers, MHI.

25. Sam H. Frank, "Air Service Combat Operations, Part 8, Operations on the St. Mihiel Salient," *Cross and Cockade Journal* 8 (Summer 1967), 75.

26. Mitchell, "The Air Service at St. Mihiel," 361–364.

27. Memorandum for Pershing, 20 August 1918, Mauer, *U.S. Air Service in World War I,* Vol. 3, 52–53.

28. Battle Orders No. 1, Air Service, in Mauer, *U.S. Air Service in World War I,* Vol. 3, 138.

29. Laurence L. Smart, *The Hawks That Guided the Guns* (np: Private Printing, 1968), 45.

30. Ibid., 46.

31. James J. Cooke, *The U.S. Air Service in the Great War, 1917–1919* (Westport, CT: Praeger, 1996), 147.

32. Frank, "Air Service Operations, Part 8," *Cross and Cockade Journal* 8 (Spring 1972), 81–82.

33. Major Lewis Brereton, "Memorandum, Lack of Cooperation Between the Artillery and Air Service, Operations of September 12 and 17th (dated 19 September 1918), in Air Force Historical Agency Archives, Maxwell Air Force Base, Montgomery, AL, File 167, 601–7.

34. Summary of Operations for 13 September 1918, in Mauer, *U.S. Air Service in World War I*, Vol. 3, 344–346.

35. Operations Report Number 17, Air Service, 1st Army, 16 September 1918, in Mauer, *U.S. Air Service in World War I*, Vol. 3, 648–655.

36. Mauer, *U.S. Air Service in World War I*, Vol. 3, 651–655.

37. William Mitchell, "Summary of Air Service Operations St. Mihiel— September 12 to 20th, 1918, in Mauer, *U.S. Air Service in World War I*, Vol. 3, 710–712.

38. Special Orders No. 120, War Department, 4 December 1918, Mitchell Papers, Carton 34.

39. Pershing to Mitchell, 16 September 1918, copy found in George Hardie Collection in the Golda Meir Memorial Library Archives, University of Wisconsin, Milwaukee, Carton 11, Scrapbook 3.

40. Paul Braim, *The Test of Battle: The American Expeditionary Force in the Meuse-Argonne Campaign* (Newark: University of Delaware Press, 1987), 96.

41. Pershing, *My Experiences,* 286–287.

42. Sam H. Frank, "Air Service Combat Operations, Part 9, Operations in the Meuse-Argonne Campaign," *Cross and Cockade Journal* 8 (Summer 1967), 174.

43. "Operations of Allied Balloons in the Saint Mihiel Offensive," Major John Pagelow, Commander, Balloons, 1st Army, 28 September 1918, in Air Force Agency Historical Archives, Entry 167.601–604, 1917–1918 (in the Frank Lahm Papers).

44. "Brief History of the Air Service, AEF," AEF History Division, 1 July 1920, Air Force Agency Historical Archives, Entry 167.401–419.

45. Unnumbered Circular, 1st Army Air Service, 21 September 1918, RG 18, Entry 767.

46. Memorandum of Understanding, Air Service, 1st Army, 23 September 1918, Mitchell Papers, Carton 45A.

47. Battle Orders No. 7, Air Service, 1st Army, 25 September 1918, RG 120, Entry 805.

48. Cooke, *Pershing and His Generals,* chapter 9. For a detailed description see Braim, *The Test of Battle.*

49. Nolan's Dictations, page 95, Nolan Papers, MHI.

50. Ibid., page 17 of Dictation number 38, Nolan Papers, MHI.

51. Frank, "Operations in the Meuse-Argonne Campaign," 182.

52. William Mitchell, "The Air Service at the Argonne-Meuse," *World's Work* (September 1919), 558–559.

53. Ibid., 559.

54. William Mitchell, *Memoirs of World War I* (New York: Random House, 1960), 265; and William Mitchell, "Greatest Raid in Aviation History," *Air Power* 4 (November 1918), 315.

55. Hugh A. Drum, "Tying Our Wings to the Ground: The Argument Against an Independent Air Service," *The Independent* 114 (14 March 1925), 287–289.

56. Orders, Headquarters AEF, Chaumont, 13 October 1918, Mitchell Papers, Carton 34.

57. Memorandum, Chief, Air Service, 1st Army, 2 October 1918, RG 120, Entry 807.

58. Ibid., AEF, 8 August 1918, RG 120, Entry 807.

59. Pershing to Kenly, 24 August 1918, RG 18, Entry 96.

60. Pershing to Adjutant General, 4 December 1918, RG 18, Entry 98.

61. Diary Entry 9 November 1918, in *The World War One Diary of Col. Frank P. Lahm* (Maxwell AFB: Historical Research Division, 1970), 145–146.

62. Memorandum to Chiefs, Air Service, 1st, 2nd Army, 26 October 1918, RG 120, Entry 805.

SIX

RETURN TO WASHINGTON

A FEW DAYS BEFORE THE ARMISTICE OF NOVEMBER 11, 1918, went into effect, John J. Pershing made a major commitment to Maréchal Ferdinand Foch and the allies. The American Expeditionary Forces would participate in an occupation force on the Rhine River. The armistice was simply an agreement that the warring sides would stop shooting while representatives of the governments negotiated a final peace. It was possible that peace talks might break down and combat begin again. As part of the November 11 agreement the allies would occupy a part of Germany on the west bank of the Rhine River. If battle began again, at least the allies would have a foothold on German soil. Pershing had tested the mettle of the German army in the Meuse-Argonne fight and knew that they could be formidable. To ensure that the American participation would be decisive, Black Jack created the 3rd U.S. Army and selected some of the best divisions of the AEF to march to the Rhine. These were the "grand old men" of the AEF, the 1st, 2nd, 42nd, 89th, and other divisions that had been in combat and now contained veteran fighters and tested leadership. There was surprisingly little grumbling, however, from those units because the young soldiers would embark on another adventure, in the enemy's homeland, and the leadership, many of whom were ambitious regulars, saw occupation duty as a new experience that would look good on records when promotion time in a peacetime army rolled around.

The AEF had learned how to use airpower during the great battles of the Western Front, and Black Jack Pershing was determined that the 3rd Army would have the aero squadrons and the balloon companies necessary to support the ground forces. With some reservations on the part of Mason Patrick, head of the AEF's Air Service, and some members of his staff, Pershing selected Brigadier General Billy Mitchell to organize the Air

Service of the 3rd Army on the Rhine River. It was a good choice because no other Air Service officer had the practical combat experience of putting together a viable, hard-hitting air force. There was always the question of how Mitchell would relate to other staff officers, however. Major General Theodore Dickman had been selected to command the new army, and his chief of staff was Brigadier General Malin Craig, who had already crossed swords with Billy Mitchell. Mitchell selected Lieutenant Colonels Lewis Brereton and John Paegelow to be his two key subordinates.[1] Brereton would orchestrate the aero squadrons, and Paegelow would work with the observation balloons.

Maintaining a viable force on the Rhine for any length of time was a tricky proposition for Pershing. Congress wanted to end the massive expenditure of funds now that the guns had fallen silent. Black Jack had the unpleasant task of ending wartime contracts with the allies, which caused a great deal of hard feelings between the AEF and other countries. There were immediate concerns that the U.S. promise to keep troops along the Rhine would mean very little as Washington sought to demobilize as quickly as possible. Initially, Army Chief of Staff Peyton Conway March, who did not get along with Pershing, sought to automatically return officers to the United States. This ran counter to the AEF's policy of selecting the best officers and aero squadrons to support the 3rd Army. In late November, March agreed to support the troops on the Rhine, suspending the automatic return of officers.[2] This helped Pershing and his staff planners, but always in the background was the congressional fiscal axe, which was ever ready to fall.

One of the most difficult problems for the entire army was how to staff positions back in the United States. The Great War had brought changes in the outlook of the U.S. armed forces, and men who had experience were needed to maintain what momentum was left after the armistice. Mitchell felt that he was in line for a major role in the development of postwar airpower, and he was deeply disappointed when Pershing sent Major General Charles T. Menoher back to the United States to take over as director of the Air Service. In his attempts to build an Air Service with the most experienced men available, Menoher soon requested that Mitchell join him.[3] Mitchell had settled into a rather fine house near Coblentz, Germany, that had once been used by the Kaiser William II's family when they had visited the region, and he was admittedly fond of the vistas offered by the house, which was situated on a high hill overlooking the Rhine.[4]

An undercurrent of dislike for Mitchell at 3rd Army Headquarters at Coblentz stemmed from difficulties with Mitchell during combat operations and from the festering Foulois-Mitchell relationship. Benjamin

Foulois had many friends around General Dickman, and they had not forgotten the eclipse of Foulois by Mitchell, even though it was Foulois who
requested that Mitchell replace him at 1st Army.[5] These very loud and sour
notes had reached the ears of Mason Patrick and John J. Pershing, and
Pershing wanted nothing to upset the smooth working of 3rd Army. To
make matters worse for Billy Mitchell, Menoher intended to use Mitchell
as the assistant to Major General William Kenly, with whom Mitchell had
had great difficulties in 1917.[6] Normally Pershing would take steps to hold
officers in Europe, especially if they were experienced in combat operations, but in the case of Mitchell he raised no objections to his immediate
return to Washington.

Before Mitchell's return to Washington, however, he was to be invested with the French Legion of Honor by no less a hero of the Great War than
Maréchal Philippe Pétain, the defender of Verdun in 1916. Mitchell also
asked Pershing for permission to visit England to confer with the chief of
the British Air Staff, General Hugh Trenchard, "to see what the result of
creating a separate branch for aeronautics has been."[7] While this did not
please Pershing, who remained skeptical of an independent Air Service, he
saw no real objection to Mitchell's stopover in England. Though Mitchell's
views of airpower on the battlefield remained fairly traditional, he had
since the end of the war become one of the loudest voices for an independent Air Service. On February 1, 1919, Billy Mitchell crossed the English
Channel and traveled to London, where, over the next few days, he met
with Trenchard for "a long talk." The long talk centered on air independence and only reinforced Mitchell's ambitions for a separate air force. On
February 4 he had dinner with Trenchard and Winston S. Churchill, himself
a supporter of air independence. By the time Mitchell left Britain on
February 17, he was a full-fledged apostle for an independent air arm that
would combine both army and navy air into one branch of the armed services.[8]

By early March 1919 Billy Mitchell was back in the United States. He
spent a brief time in Washington preparing for his new position and then
went to Rochester to see his family. Caroline Stoddard Mitchell had had an
interesting and productive war effort herself. After a stint as an ambulance
driver and repair mechanic she took a quick course as an assistant nurse at
the Rochester General Hospital, earning her certificate. She worked in the
operating room and as an assistant on the wards during the terrible influenza epidemic of 1918–1919. She was an independent and accomplished
woman who had not seen her husband for two years. In the spring of 1919
she announced that she was pregnant with their third child, who was born
on January 20, 1920, and was named John Lendrum Mitchell II.[9] Home life

with Mitchell seemed at first to go well as he regaled the family with his war stories and with plans to move the family as soon as possible back to Washington, where their child would be born. But there were nagging problems with Billy Mitchell in that he seemed greatly distracted by his dedication to achieving an independent air arm, and he was very quick to criticize and lash out at General Menoher. It distressed Caroline that William Mitchell was drinking heavily, and he did not seem to be willing to curb expenses, especially when it came to horses and horse clubs in Washington and Virginia. The drinking was also noticed by his sister Harriet and by his mother, who knew well the charges of "habitual intemperance" leveled at Mitchell's father by his first wife. There were dark clouds on the horizon for the William Mitchell family.

Of course it was possible to rationalize Mitchell's behavior by citing his war experiences, and the stresses and tensions of commanding a brigade, then an army, and then an Army Group. Airpower was new to the battlefield, and Mitchell had little guidance, except from the British and especially the French, as to how to organize his forces and how to employ them against a very battle-tested enemy in the skies. Mitchell had become in France very much the "intrepid airman" the press loved to write about. But this was not the genial and highly praised young officer who left Washington for France in early 1917. Once back in Washington, and with Caroline going through a difficult pregnancy, Mitchell sought out more and more allies in Congress, and he surrounded himself with subordinate officers whom he had commanded in France and who idealized their commander. His titles were impressive: Director of Military Aeronautics and Chief of the Training and Operations section. He brought his old friend Colonel Thomas D. Milling to Washington as his assistant. Lieutenant Colonels William Sherman, Lewis Brereton, Harold Hartney, and Colonel Charles DeF. Chandler (chief, Air Service Balloons and Airships) rounded out this close-knit group that mirrored Mitchell's views.[10] There was almost a fortress mentality surrounding Mitchell and his subordinates, and much of their ire was focused on Major General Menoher, who sported no flyer's wings on his uniform—Menoher was an artilleryman.

But General Menoher had something very important: the patronage and goodwill of General of the Armies Black Jack Pershing. It was Pershing who had made sure Menoher would keep his two stars while other wartime generals were reverting to lower ranks. Brigadier General Benjamin Foulois, for example, found himself replacing his single star with the gold leaves of an army major. Menoher was a solid officer, an 1886 graduate of West Point, and a classmate of John J. Pershing. His career in the artillery was good enough to earn him admission to the army's new War College.

Menoher had served with Pershing in the Philippines, and he had four very productive years on the General Staff. When Pershing was sent to France in the summer of 1917 he brought Menoher to France despite the fact that he was on orders to command the artillery brigade of the 82nd Infantry Division in training at Camp Gordon, Georgia. In mid-December 1917 Black Jack gave him command of the 42nd, the Rainbow Division, a choice assignment, and his combat command was one of the longest and most successful in the AEF. His chief of staff, and later one of his brigade commanders, was young Brigadier General Douglas A. MacArthur, the same man who had once had a crush on Harriet Mitchell back in Milwaukee.[11] Determined to keep Menoher as a general officer in the postwar, very lean, small Regular Army, Pershing had assigned him to the post of chief of Air Service despite his nonflying status; in fact, Menoher had never flown in a plane or made an ascension in an observation balloon. But Menoher believed in the Pershing orthodoxy, with airpower being on the battlefield to see that the infantryman with rifle and bayonet won the fight. In Charles T. Menoher's mind there could be no such thing as an independent air force.

Prior to Mitchell's departure from Europe and his several meetings with Hugh Trenchard, his tactical views had been largely orthodox. In early January he wrote a short paper entitled "Tactical Application of Military Aeronautics," which summarized his battlefield experiences, and contained within the paper was a section on rigid airships, or dirigibles. Of importance was Mitchell's view that airships could be used to engage the navies of the world. His mind had begun to turn toward the role of the navy in future wars, and his response to those seapower advocates was that air had changed the war-fighting dynamic.[12] Though his early views on the viability of the battleship navy were vague, they would resonate with Mitchell over the next six years in Washington. Menoher, in contrast, did not want a subordinate who looked outside of the army for new areas of new combat. Like many high-ranking officers he realized that the United States Congress was going to be tight with the purse strings. The armed services would be, from 1919 on, in competition for what few dollars Congress saw fit to allocate for defense. In a memorandum to Mitchell in early March 1919, Menoher warned that his office could not be enlarged by a huge staff; Mitchell was to maintain a lean office and coordinate with other branches of the Air Service and the army. Within set limits, however, Menoher, who was not yet comfortable with his new job, would allow the very experienced Mitchell to develop plans and to orchestrate training and oversee new research.[13]

Had Mitchell not been so antagonistic to the nonaviator Menoher, he might have found a willing ally and a steadying force. Menoher might have been a true believer in Pershing's cult of the rifle, but he was also a highly intelligent officer who could examine for himself the great changes that air-power had brought to the Western Front. In late February he was invited to speak before the Aero Club of America at its annual meeting at the Waldorf Hotel in New York. Menoher told his audience that in future wars, with wireless communications, generals would direct their troops from the air.[14] A month later he surprised many Regular Army officers of the General Staff when he addressed the Aeronautical Exposition in Madison Square Garden in New York and called for the establishment of an air academy similar to the Military Academy at West Point.[15] When interviewed for the *New York Times* in April, General Menoher stated that every agency of government that had an interest in air development had to cooperate due to the special needs of the service.[16] He stopped short of calling for a unified Air Service for the army, navy, post office, and other agencies, but the implications were there.

As chief of the Air Service, Menoher was in constant demand as a speaker, especially by influential organizations devoted to the growth of American aviation. But while he was building public support for army aviation, his office in Washington was not running smoothly. Colonel Oscar Westover, Menoher's chief assistant, had more dealings with Mitchell and his Training and Operations Group on a day-to-day basis than did Menoher, and he was becoming more frustrated by Mitchell and by Milling. Westover recognized that Mitchell had brought into the Air Service office men with whom he served with, and that these officers, from Milling on down, had adopted Mitchell's views. In Westover's view, when Menoher's office issued a directive Mitchell's men did nothing. "Something is certainly wrong," Westover told his chief, "[Mitchell] . . . is seizing upon every opportunity to press a point which will gradually cause the adoption of schemes or plans recommended by him." Colonel Westover then recommended to Menoher that Mitchell be removed from his position and that Training and Operations be given to a more responsive, less combative, officer of the Air Service.[17]

A few of Mitchell's wartime commanders, however, continued to think highly of his abilities. In late January 1919, Major General Mason Patrick filed a glowing efficiency report on Billy Mitchell that read, in part, "I believe General Mitchell to be a very efficient officer. His personality is pleasing, he is likeable, he possesses exceptional qualifications for leadership, and . . . he thinks rapidly and acts quickly, sometimes a little too hastily. He is opinionated but I have usually found him properly subordinate and ready to obey orders." This would not have been the opinion of Benjamin

Foulois or Frank P. Lahm, nor would Hugh Drum, Dennis Nolan, or Malin Craig have agreed with Patrick's assessment. General Patrick did include a possible warning for Pershing: "He has some tendency to act on his own initiative; it is not meant that this is a fault, as it is frequently a virtue, but there have been a few times when it has been uncertain just where he was or what he was doing."[18]

Menoher was now in a quandary. There were very distinct factions in the chief's office in Washington, and antagonism was growing between them. This came at a time when Congress was debating the future budget of the armed forces, and General Pershing had convened a board under Major General Theodore Dickman to discuss Air Service matters. A fractured Air Service in Washington could derail fiscal legislation. The military thing to do was to relieve Mitchell and assign him somewhere else in the United States or in its territories, but Mitchell had powerful friends in the House of Representatives and the Senate. His social connections, growing out of his love of fine horses, included the family of the powerful senator from New York, James W. Wadsworth Jr. Wadsworth and Mitchell met often on Capitol Hill to discuss military matters, and they saw each other at horse shows, where the senator's daughter was very friendly with William and Caroline Mitchell.[19] In addition, there were still a few old congressional hands who remembered the Democrat from Wisconsin, Senator John Lendrum Mitchell. This was an unappealing, and perhaps unsolvable, set of problems for Menoher, the professional soldier, who had few friends on Capitol Hill.

Mitchell was becoming more and more antagonistic toward Menoher and his staff officers. It was noted that Mitchell spent much time away from his office, and other officers complained that they could never find anyone in the office to answer questions. Communications from Mitchell's office were, despite Menoher's express instructions, erratic, and Mitchell ignored guidance from Menoher about writing or speaking without his chief's approval.[20] Constantly sniping at Menoher, Mitchell told New York Congressman Fiorella LaGuardia's House Subcommittee of the Committee on Military Affairs, "The great trouble has been and still is to some extent that officers with no air training are placed virtually in command of air forces for which they are in no way suited."[21] As noted earlier, Menoher did not wear the wings of an aviator, and Mitchell's statements were interpreted as a slap at his superior officer, a breach of military courtesy and discipline. LaGuardia, a Great War aviator himself and a Billy Mitchell supporter, pushed Mitchell to speculate as to what aviation needed. Quick to respond, General Mitchell called for a unified air service with a cabinet-level position devoted to air matters. When asked if this was the navy's

position, Mitchell responded that he believed a vast majority of navy aviators favored a unified service, and this brought howls of protest from the
Navy Department to Newton B. Baker, Secretary of War, and to Menoher.

Mitchell was quickly becoming a thorn in the side of the War
Department and the army over his frequent speeches and writings calling
for a separate air service. Because of Mitchell's political connections and
his general public audience, Menoher and others tried to bring him into all
discussions. In August 1919 an army board was convened to look into the
practicality of an independent air arm, and Mitchell was asked to give his
opinions. Much to the aggravation of Menoher and other general officers,
Mitchell proceeded to lecture them as if they were obtuse cadets. Mitchell
requested that his statements be taken verbatim by Menoher's stenographer,
but that was refused. He then engaged one of the generals in a debate over
the possibility that airpower alone could win a battle or a war. Major
General William G. Hann, an 1889 West Point graduate, successful commander of the 32nd Division in the war, and a favorite of Black Jack
Pershing, argued that it was the infantry that won fights. Mitchell countered
by telling Hann that he was like so many other generals who knew nothing
about the Air Service, and, because of their branch of service in the infantry
or artillery, would make no effort to learn anything about the air. The meeting with Mitchell broke up with a great deal of anger and frustration, and
when Mitchell wrote his notes a few days later he stated that "the whole
hearing impressed on me more than ever that, under the control of the
army, it will be impossible to develop an Air Service."[22]

After Mitchell's outspoken criticism of American air efforts in general
and the navy's attitudes in particular, Secretary of War Newton B. Baker
was furious. Being pressed by the Navy Department, Baker required
Mitchell to answer questions about his sources when he stated that navy
fliers favored an independent air arm, but that "[navy aviators] hesitate to
express their opinions because they are all junior officers and because the
senior officers are against it largely, I believe, from lack of familiarity with
the subject." Mitchell's response was vague and glib, almost ignoring specific requests from his civilian superior. Baker and the Navy Department
had had enough of Mitchell's statements, which, they pointed out, had no
specific references or documented sources.[23] Truth be told, Billy Mitchell
was being unfair about the navy, which had a number of senior officers
who were airpower advocates and who could have been Mitchell allies had
he not been so intemperate with his remarks. By the end of 1919 and well
into 1920 Mitchell increased his attacks on the navy, and many in the Navy
Department believed that Mitchell wanted control of all air assets regardless of departmental responsibilities.

While Mitchell was stirring the pot of controversy, his home life was deteriorating. Caroline's pregnancy was a difficult one, and doctors reported that there was a chance that natural childbirth might not be possible. She remained at home a great deal because of extreme discomfort. Mitchell's mother and his sister visited Washington to assist Caroline as much as they could. Her own mother traveled several times from Rochester to the nation's capital to see her daughter. While this situation was of great concern to both families, Mitchell was spending money on various things such as club memberships, horses, and riding clothes imported from Britain, which the Mitchells could not afford. In 1919 and 1920 Mitchell was a member of the Rock Creek Hunt Club and several horse clubs in Washington, and in 1920 his eleven-year-old daughter Harriet, riding a horse named Sir Dixon from the "Mitchell Stables," won first-place blue ribbons for show and for jumping at the National Capital Horse Show.[24] The so-called Mitchell Stables and membership dues and entry fees, not to mention entertainment, plunged Mitchell deeper into debt.

In late 1919 Mitchell borrowed $2,000 from a Washington bank to cover his many expenses, and he planned to borrow even more to see him through 1920. Increasing bank debt was risky for an army officer, and, knowing how the army frowned on indebtedness, he asked his mother for a loan of $1,000. "I have cut down expenses just as much as I can and the only way I can cut down any more is to get out of Washington or leave the service and make more money," he wrote to his mother.[25] As he asked for assistance he wrote something very revealing, saying, "I am practically the only one that can bring about a betterment of our national defense projects at this time. Leaving here now would have that effect. Also this [a unified Air Service] is the one project that I most want to leave behind as an accomplished fact when I go."[26] Mitchell's private correspondence with his "Dear Mummy" over a period of several months in late 1919 and into early 1920 reveals his self-image as a single warrior against the entrenched establishment of the armed services. In December he wrote his mother, who was preparing to leave for Europe to visit and decorate the grave of John Lendrum Mitchell Jr., that "his work," the creation of an independent, unified Air Service, would be completed in a decade due to his efforts alone.[27] He overlooked the work of men like Benjamin Foulois, who was very much in evidence on Capitol Hill and among industrialists concerned with aircraft production.[28]

By Christmas 1919 Billy Mitchell was in hot water again with Newton Baker, General Menoher, and the entire General Staff. On October 26, 1919, Congress refused to appropriate $15 million for the purchase of American-made airplanes for the Army Air Service. Baker, Menoher, and

Mitchell had gone to Capitol Hill to lobby for passage, but Congress was in no mood to vote extraordinary funds in the wake of the vast expenditures required to fight World War I. Mitchell was justifiably angry, and he sent a memorandum to Newton Baker pointing out what the failure would mean. If the United States could not buy its own military aircraft, then the nation, in a time of crisis, would have to rely on the European states, assuming that they would be friendly. Then the Europeans, Mitchell wrote, would, "absolutely control our Air Service and therefore what I believe will be the most important single element of our national defense." In a dire prediction, Mitchell told Baker, "The factories and the airplane industry in this Country are merely waiting and holding to get orders, which, if they do not get, will force them to go out of business."[29] Mitchell was certainly correct to be angry at the Congress, and he had personally experienced problems caused during the war because the United States had to rely on the French and the British. These dire warnings were also raised by Baker and Menoher; Mitchell, however, had gravitated to a position that only government contracts and the needs of the military provided support for the American aviation industry. Billy Mitchell had begun to view, and would continue to advocate, the central government as the arbiter of the aircraft industry. Baker read Mitchell's memorandum, which said little that was substantially new, and had his assistant, Colonel James A. Blair Jr., reply to Mitchell, saying that "[Secretary Baker] did not see how anything could be done immediately on this matter—until the whole subject of aviation had been settled."[30] That would have ended the disappointment felt by all had Mitchell been content to stop there.

Right before Christmas, the Manufacturers Aircraft Association of New York sent to editors of various magazines a copy of an article by Mitchell entitled "Why We Need a Department of the Air." The article, for which Mitchell had been paid, began with the thesis that the surface navy was rendered irrelevant in the defense of the United States with the coming of airpower; the "air will prevail over the water in a very short time, particularly as radio communication between aircraft and submarines has become an accomplished fact." He then went on to lobby for upcoming legislation, for a unified Air Service, and for a cabinet-level Department of the Air that would cut across military and civilian lines.[31] While none of this was new as far as Mitchell was concerned, it did place Mitchell squarely on the side of manufacturers as a paid spokesman—and he did this without telling Menoher, his boss, and Baker, his civilian superior, anything in advance. This was the type of Mitchellesque writing that was bound to bring Secretary of the Navy Josephus Daniels and the entire Navy Department after Baker and Menoher with fire in their eyes, and it did just that.

To make matters even worse, after Mitchell's testimony before the LaGuardia Committee, Josephus Daniels had protested in writing to the committee. Mitchell was called back in late December by LaGuardia and asked several times by various congressmen if he stood by his statements that junior officers of the navy who were aviators favored a unified service. Mitchell continually and with no hesitation answered "yes," and he even took the opportunity to elaborate on them. But when asked for names, Mitchell refused, stating that the men might be subject to reprisals by the Navy Department.[32] Daniels was again furious over Mitchell's obvious attacks on the Navy Department, and the latest article would only increase his ire.

The Mitchell-Navy Department conflict could be boiled down to two main points: First, Mitchell was determined to see a unified Air Service with a cabinet-level position dedicated to air matters; and second, there were severe battles between the army and the navy over shrinking financial support from Congress, and Mitchell's involvement didn't help matters. The navy assigned their best spokesmen to lobby for increased funding. The eloquent and forceful Assistant Secretary of the Navy, Franklin Delano Roosevelt, was sent on the road to address civic groups about the need for more funding. In late January 1920 he told the Brooklyn, New York, Chamber of Commerce that to keep the United States as a premier naval power it was possible that the nation might have to spend $1 billion per year. When greeted with stunned silence, he responded that the cost was necessary to keep the navy strong.[33] Roosevelt was also one of Mitchell's greatest opponents, calling his ideas for a unified Air Service a "pernicious doctrine." The army's Dickman Board had determined that air independence was justifiable only if it could be shown that air would be the decisive arm in war. The board did not believe it could be shown to be so, and Roosevelt echoed those sentiments loudly.[34]

Josephus Daniels also took to the field to publicly combat Mitchell's ideas. On March 6, 1920, Daniels testified before the House Naval Committee, and he was fighting mad. During his time with the members of Congress he attacked Mitchell by saying that airpower had not rendered the battleship obsolete, and, in fact, aviation was "essentially an integral part of the fleet." If an independent Air Service was created it would be doomed to failure, and it would be impossible to create a cabinet-level position that represented both military and civilian aviation. The two sectors simply had very different needs and objectives. In a direct slap at Mitchell, who believed the government had a major role to play in all aviation, Daniels stated that commercial aviation should be encouraged, of course, but not at taxpayer expense or with government oversight. Daniels ended his testimony by saying, "There are some who believe that capital ships will soon be at the mercy of aircraft. It is not believed, however, that responsible officials of any admiralty charged with the organization and operation of the first line of defense accepts even approximately any such view."[35]

Mitchell had been loudly proclaiming and writing since 1919 that airpower made the expensive battleship an obsolete weapon of war. As far as Mitchell was concerned the battle lines were drawn, and he attacked quickly. In speeches and articles he challenged the navy to show that airpower could not sink a modern battleship. As the war of words heated up, Secretary of War Newton Baker, who was very skeptical about Mitchell's calls for an independent Air Service, directed that all officers, especially those of the Air Service, be aware that testimony before Congress, articles, and statements to the press had to be cleared with him. Baker was openly upset when Mitchell continued to attack the navy, and he directed Menoher to make certain all Air Service officers understood his guidance in respect to public pronouncements.[36] Menoher did not have a firm grip on the Air Service, and he certainly did not know about Mitchell's ongoing plans with Colonel Thurman H. Bane, who was chief of the Army Air Service's Engineering Division at McCook Field near Dayton, Ohio.

In September 1919 Mitchell evidently surfaced an idea with Bane about hitting a ship with bombs. Bane, who was perfectly willing to cooperate with Mitchell's embryonic schemes, warned him that if the question of proper bombs for a test rested with the United States Army's Ordnance Corps, nothing would ever be done. Development should, according to Bane, rest with the Air Service, and properly with his division at McCook Field.[37] Mitchell kept Bane updated as to the progress, or rather the lack of it, in getting Congress to devote more funds to the armed services. He also asked him to visit Orville Wright to enlist the aviation pioneer to testify before Congress in favor of a unified Air Service.[38] Bane's meeting with Wright was not satisfactory at all because Wright did not really have an opinion, though he felt that in the long run, given the nature of research and development, the present system might be best. Bane subsequently advised Mitchell not to ask Wright to get anywhere near Congress, or the Navy Department for that matter.[39] Mitchell's relationship with Colonel Bane fluctuated wildly as Mitchell became more and more interested in bombing to prove his point with the navy. He lashed out at Bane in March 1920, accusing him of failing to procure a proper bomb. Bane had stated the very obvious fact that the Engineering Section was critically short of money, an ailment that afflicted all services, but Mitchell would hear none of it and added an unkind cut, stating, "There are several reasons in my opinion why your Engineering Section has been slow to get material out. In the first place, the Engineering did nothing during the war."[40] When Colonel Bane was reduced from his wartime rank to major, Mitchell dismissed it, writing, "So far as I am concerned, whether I am made Assistant Director or Brigadier General is actually a matter of minor importance with me, because the work is so absorbingly interesting. . . . There is no use crying over spilled milk."[41] The gap between Mitchell's starred position and Bane's new gold major's leaves, however, was great. Mitchell had given Bane the impression that he was a man of means with his horses, club memberships, and imported British riding clothes. This arrogance directed against a suffering Thurman Bane was not justified given the different military circumstances of the two officers. Every relationship he had—from hard-pressed subordinates to his family—appeared to suffer as Mitchell became convinced that he could show the navy what airpower could do—if he could only get a battleship to sink.

Mitchell knew that the navy was conducting experiments with the obsolete battleship *Indiana* by dropping dummy bombs from aircraft and then setting off pre-placed charges to gauge the effect of the explosions on the ship.[42] When Mitchell asked Menoher to help secure a ship, his chief was aghast. Mitchell had been observing the late-summer *Indiana* operations and taking pictures of what the navy was doing. In September Mitchell stepped up his requests for a ship, writing to Menoher, "We must at all costs obtain the battleship to attack and the necessary bombs . . . to make the test a thorough and complete one. . . . An act past [*sic*] by Congress last year specifically charges the Army Air Service with coast defense and this necessarily implies the attack of hostile vessels. [A battleship] can be furnished by the Navy in precisely the same way in which we furnish the Navy so much equipment."[43] Mitchell's plan made sense; it would do little good to wait until a possible enemy was within striking distance of the U.S. coastline and then wonder how it was to be fought and defeated.

On November 1, 1920, after repeated "hits" by naval aviation and pre-placed charges, the *Indiana* sunk in the waters off the Virginia coast. Mitchell was beside himself with glee as the great old ship went down: "There is no question whatever about the fact that the ship on the surface of the water is gone, and that the aircraft has taken its place."[44] Respected naval officers had reached virtually the same conclusions as Mitchell, and they could have been allies had Mitchell used them skillfully. One such officer was Rear Admiral William Freeland Fullam, a New Yorker who had graduated from Annapolis in the class of 1877. Fullam had written a number of works and had been the superintendent of the Naval Academy in 1914 and 1915. During the Great War Fullam had commanded the Patrol Forces of the United States Pacific Fleet before he retired in 1919. After the

Indiana bombings Mitchell sent a full report to Fullam, who was staying in Philadelphia, Pennsylvania. Mitchell described in detail what had happened to the *Indiana,* which must have interested Fullam, but then he went on to point out what was needed as far as aircraft carriers were concerned, which Mitchell knew very little about. He proposed taking three captured German commercial ships—the *Agamemnon,* the *Von Steuben,* and the huge *Leviathan*—and converting them to carriers at a total cost of $12 million. They could be faster than any battleship and could, in Mitchell's scheme, carry well over one hundred pursuit planes to defeat enemy aircraft. "In other words, neither coast defense gun nor artillery on a warship need fire a shot if an Air Force is properly organized for offensive work."[45]

What Mitchell did not say was that the Operations Division of Mitchell's office already had a comprehensive combat plan for attack against enemy battleships that had been prepared by Major W. G. Kilner, chief of the Operations Division. The plan could easily be converted to a training operation once Mitchell had secured his battleship and the green light to attack it.[46] In February 1921 Josephus Daniels reported the *Indiana* results to Newton Baker, telling Baker that the navy was ready to progress to a more modern battleship by spring. Daniels then suggested that the army participate in the air attacks on the unspecified modern battleship. To circumvent Mitchell's constant harping on the navy, Daniels suggested that the whole matter be referred to the Joint Board of the Army and Navy for recommendations on how to proceed.[47] On the same day that Daniels was writing to Baker, Mitchell wrote to Bane that "during the coming year we must concentrate on bombardment above all things."[48]

Mitchell continued to inform Admiral Fullam of his plans, with the hope that Fullam would communicate with other like-minded officers in the navy. The navy, however, was moving toward the establishment of a Bureau of Aeronautics, and Mitchell now had the tough Admiral William A. Moffett, director of Naval Aviation, as an enemy. Moffett was more than willing to confront Mitchell and to try to counter his continual appearances before Congress.[49] While Mitchell was marching up to Capitol Hill as often as he could, he continued to plan for his bombing tests, which he hoped would take place in the spring or early summer of 1921. Everyone around Mitchell was being drawn into the upcoming tests. Lieutenant Colonel James E. Fechet, chief of Training and Operations, was constantly looking for ideas that would deliver larger and better bombs on target.[50] Mitchell was fully determined that Army Air Service bombers would drop live bombs on the target. Josephus Daniels was aware that Mitchell planned to upstage the navy when bombing began, and to forestall a performance by Mitchell and his Langley Field, Virginia, aviators he informed Baker that since the navy would provide the ship, it would be in order for the army to submit its plans for the upcoming operation. Daniels asked Baker to phrase the order to address their requirements and plans to the commander-in-chief of the Atlantic Fleet so that "suitable arrangements [could] be made for the Army Air Service participation under his direction."[51] That would certainly be a red flag for Mitchell and his aviators.

Mitchell's allies in Congress were active on his behalf. When, on May 4, 1921, Acting Secretary of the Navy Franklin D. Roosevelt briefed interested members of Congress, Senator Roy G. Fitzgerald of Ohio quickly informed Mitchell that the presentation was, "tiresome, although many of the pictures were interesting." The congressman then advised Mitchell as to how to make an Army Air Service presentation more interesting, especially if it showed "the helpless condition of great ships against attack from the air."[52] Admiral Fullam entered into the fight, telling the powerful Senator Niles Poindexter that "*Battleships alone* [Fullam's emphasis] will not suffice."[53] The whole army-navy fight over airpower was heating up, and, frankly, Menoher had lost control of the battle. The fight was between Mitchell and his allies and Roosevelt, Daniels, and their friends.

While Mitchell's campaign for an independent Air Service and his fight against the battleship navy was gaining momentum, his home life was breaking down. Mitchell was a man possessed, spending more and more time in his office, on Capitol Hill, and with his aviators at Langley Field. Caroline was becoming more concerned with her husband's behavior, his spending on stables and horses, and his increasing drinking. There were continual verbal confrontations between Caroline and Mitchell, and in a few months Mitchell would leave his home for officer's quarters in Washington and at Langley Field. Menoher was certainly concerned that Mitchell's office, only a few doors from his, was usually unmanned, with no officer present to answer a question or take a message. On the verge of his most spectacular operation since the St. Mihiel operation in September 1918, Mitchell was the center of attraction, but it was costing him dearly with his wife and children.

NOTES

1. HQ, 3rd U.S. Army, Roster of Officers, Christmas 1918, in the William Mitchell Papers, Library of Congress, Washington, DC, Carton 50. (Hereafter cited as Mitchell Papers.)

2. March to Pershing, 20 November 1918, in Records Group 18, U.S. Air Forces, National Archives, Washington, DC, Entry 96. (Hereafter cited as RG 18.)

3. William Mitchell, *Memoirs of World War I* (New York: Random House, 1960), 305–306.

4. Mitchell to Mother, 13 December 1918, in the George Hardie Collection, the Golda Meir Memorial Library Archives, University of Wisconsin, Milwaukee, Carton 11, Scrapbook 3. (Hereafter cited as the Hardie Collection.)

5. Diary Entry, 24 March 1919, in The Mason Patrick Diaries, United States Air Force Academy Archives, Colorado Springs, CO.

6. Menoher to Pershing, 7 January 1918, RG 18, Entry 96.

7. Pershing to March, 12 January 1919, RG 18, Entry 96. General Menoher agreed with Pershing's position as long as Mitchell did not prolong his stay in Britain. Menoher to Pershing, 15 January 1919, RG 18, Entry 96.

8. Diary Entries for 1, 2, 4, and 8 February 1919, Mitchell Papers, Carton 4.

9. Biographical information provided from the files of the Alumnae and Alumni Society of Vassar College, Poughkeepsie, New York.

10. Diary Entry, 11 March 1919, Mitchell Papers, Carton 4.

11. For more see James J. Cooke, *The Rainbow Division in the Great War, 1917–1919* (Westport, CT: Praeger Publishing, 1994).

12. William Mitchell, Mimeograph, "Tactical Application of Military Aeronautics," in U.S. Air Force Historical Agency Archives, Maxwell AFB, AL, File 167.4–1.

13. Memorandum by Menoher, 8 March 1919, Mitchell Papers, Carton 7.

14. *New York Times,* 20 February 1919, 11.

15. Ibid., 15 March 1919, 15.

16. "Active Aggression Policy for Army Air Service," Ibid., 6 April 1919, Sec. IV, 7.

17. Westover to Menoher, 26 June 1919, Mitchell Papers, Carton 7.

18. Report filed by Major General Mason Patrick, 17 January 1919, in Records Group 120, Records of the AEF, Adjutant General's Files, National Archives, Washington, DC, Carton 2264.

19. Martin L. Fausold, *James W. Wadsworth, Jr: The Gentleman from New York* (Syracuse: Syracuse University Press, 1975), 149–150.

20. Walter Frank to Mason Patrick, 28 January 1928, in the Benjamin Foulois Papers, Library of Congress, Washington, DC, Carton 5. (Hereafter cited as the Foulois Papers.)

21. Transcript of Mitchell's Statement, 20 December 1919, Mitchell Papers, Carton 32.

22. Mitchell's Private Notes, 16 August 1919, Mitchell Papers, Carton 32.

23. Memorandum for Mitchell from Colonel Oscar Westover, 22 December 1919, Mitchell Papers, Carton 7.

24. Clippings, 1919–1920, in the Hardie Collection, Carton 10, Folder 8.

25. Mitchell to Mother, 20 February 1920, Mitchell Papers, Carton 19B.

26. Ibid.

27. Ibid., 18 December 1919, Mitchell Papers, Carton 19B.

28. Foulois to Friend, 6 September 1919, Foulois Papers, Carton 5.

29. Mitchell to Baker, 29 October 1919, Mitchell Papers, Carton 7.

30. Blair to Mitchell, 30 October 1919, Mitchell Papers, Carton 7.

31. Article by Mitchell, Dated 21 December 1919, Mitchell Papers, Carton 25.

32. Transcript of Mitchell's Testimony, 20 December 1919, Mitchell Papers, Carton 34.

33. *New York Times,* 1 February 1920, 20. Also see Franklin D. Roosevelt, "Why Naval Aviation Won," *U.S. Air Services* (July 1919), 7–9.

34. Alfred F. Hurley, *Billy Mitchell: Crusader for Air Power* (Bloomington: Indiana University Press, 1975), 46–47.

35. *New York Times,* 7 March 1920, 1, 12.

36. Memorandum by Menoher, 16 June 1920, citing Baker's Memo of 9 April 1920, Mitchell Papers, Carton 8.

37. Bane to Mitchell, 3 September 1919, Hardie Collection, Carton 6.

38. Mitchell to Bane, 9 September 1919, Hardie Collection, Carton 6.

39. Bane to Mitchell, 13 September 1919, Hardie Collection, Carton 6.

40. Mitchell to Bane, 20 March 1920, Mitchell Papers, Carton 7.

41. Mitchell to Bane, 2 July 1920, Mitchell Papers, Carton 7.

42. William F. Trimble, *Admiral William A. Moffett: Architect of Naval Aviation* (Washington: Smithsonian Press, 1994), 74–75.

43. Mitchell to Menoher, 27 November 1920, Mitchell Papers, Carton 8.

44. Mitchell to Major A. H. Hobley, McCook Field, 2 November 1920, Mitchell Papers, Carton 7.

45. Mitchell to Fullam, 30 December 1920, in the William F. Fullam Papers, Library of Congress, Washington, DC, Carton 5. (Hereafter cited as the Fullam Papers.)

46. "Plan for Operations for Bombardment by Airplane of Single Battleship," 11 December 1920, Mitchell Papers, Carton 8.

47. Daniels to Baker, 7 February 1921, Mitchell Papers, Carton 9.

48. Mitchell to Bane, 7 February 1921, Mitchell Papers, Carton 7.

49. Trimble, *Admiral Moffett,* 74–75, 77–79.

50. Fechet to Mitchell, 17 February 1921, Mitchell Papers, Carton 9.

51. Daniels to Baker, 5 March 1921, Mitchell Papers, Carton 9.

52. Fitzgerald to Mitchell, 9 May 1921, Mitchell Papers, Carton 9.

53. Fullam to Poindexter, 21 March 1921, Fullam Papers, Carton 6.

SEVEN

THE *OSTFRIESLAND*

B Y THE SPRING OF 1921 MITCHELL WAS A WELL-KNOWN figure in the news. His upcoming bombing of a navy-owned battleship was covered by a press anxious for a story with drama, a story whose outcome was very much in doubt. And Mitchell played it for all it was worth, angering his boss, General Charles Menoher, and Colonel William F. Pearson, Menoher's executive officer and brother-in-law. General Mason Patrick, who had commanded the Air Service in the American Expeditionary Forces in France, watched with dismay as Mitchell seemed to be conducting his own operation, separate and distinct from the mission of the peace-time Air Service. Patrick felt that Mitchell had "really little else to do [but spend] a large part of his time in political activities, agitating for a separate Air Service, and continuing the aircraft vs. surface vessels controversy with the Navy."[1] Menoher had no control over Mitchell, and this was not lost on Black Jack Pershing, who was scheduled to become the army's chief of staff during the summer of 1921, when General Peyton C. March retired from active service.

Before bombs fell on battleships from aircraft the war between Mitchell and the Navy Department was an overheated battle of words. Editorials supporting Mitchell appeared in the *New York Times* and the New York *Tribune,* and the magazine *Aerial Age Weekly* trumpeted Mitchell as the air hero of the age.[2] Retired Admiral William F. Fullam sent out as a press release a letter he wrote to a member of Congress thanking him for his support of naval appropriations, saying, "But the Navy Department has not put the cards on the table. . . . Admitting that the battleship is not yet dead—that it is the backbone of the Navy—my contention is that the backbone will be powerless if it is unprotected and unsupported by submarine and air forces."[3] The well-read magazine *The Independent* came down on

the side of Mitchell, citing the great cost of building and maintaining a battleship, which cost about $40 million, as compared to a bomber, which had
a price tag of $40 thousand. It just made sense, argued the writer, that the
Mitchell bombing trials be conducted fairly and in public view.[4] In April
1921 Mitchell published an article in *World's Work* entitled "Has the
Airplane Made the Battleship Obsolete?" Of course, his well-argued thesis
was that the airplane had indeed relegated the battleship to the same historical trash heap as the smoothbore musket and the pike.[5]

Mitchell was pushing his subordinates as hard as possible. He wrote to
Thurman H. Bane, chief of the Engineering Division at McCook Field, that
he had to double his efforts to get bombing aircraft to Langley Field. "I
want to impress upon you the necessity of getting those ships out and getting them to Langley in the *shortest possible time* [Mitchell's emphasis].
This is the most important thing there is today."[6] The harder Mitchell
worked to prepare for the upcoming bombing runs, the more his personal
and professional life suffered. In May newspaper articles appeared citing
the strain between Mitchell and Menoher, which seemed to be more than
just a clash of personalities. Not being fully informed as to what his head-
strong subordinate was doing, Menoher was embarrassed several times by
not being able to answer questions posed by members of the General Staff
or by the press. Mitchell's political contacts rallied to his side, and
Congressman Hubert F. Fisher of Tennessee wrote a typical letter, saying,
"I do not see how it would be possible for the Secretary of War to do anything else than to stand behind the one flying general of the army who has
done so much for aviation. . . . I have talked with your several friends on
the Hill and they all quite agree with me."[7] It was not only Mitchell against
a battleship and the navy, it was now Mitchell against Menoher.

At the same time, Mitchell's marriage to Caroline fell apart. His battle
with the navy kept him away from home most of late 1920 and well into
1921, and the confrontations between Billy and Caroline became more rancorous. Caroline could see clearly that her husband was almost out of control, and his drinking was a matter of great concern. There were unsubstan-
tiated rumors that Mitchell, still a handsome man, the debonair flyer, the
general-hero of the Great War, might be involved with other women, and
Caroline had confronted him with the stories.[8] On June 14, 1921, after a
bitter fight, Mitchell packed his bags and left the house,[9] and Caroline took
the children and returned to Rochester to await events. Mitchell had
become absorbed with the bombing runs and with his open conflict with
Menoher, and he told his mother later in the summer that he had not realized the strain on himself until he actually left the house and was installed
in officers' quarters in Washington and at Langley Field.[10]

On June 2, 1921, Navy Aviation attacked the captured German submarine U-117 and sunk it, and on June 5 the Army Air Service went after the German Destroyer G-102. Mitchell's plan was to simulate a full air attack showing how the Air Service would operate against enemy ships. Eighteen pursuit planes came in first to take care of enemy aviation and anti-aircraft weapons. Other bombers would follow, hitting the heavier ship's auxiliary vessels, which, in time of war, would be carrying fuel, food, and ammunition. Once those ships were destroyed the heavier bombers would attack the G-102 full force. The bombing worked as planned, breaking the captured ship in two and sending her to the bottom. The operation had been a great training exercise, well planned by Mitchell's subordinates at Langley Field, and was more complex than the observers on the nearby vessels could have imagined. Mitchell had finished the first round of his fight with the navy in fine form, and back at Langley Field he and his flyers celebrated.

By cajoling and browbeating subordinates like the harried Thurman Bane, Mitchell had obtained a small supply of very lethal 300-pound bombs for the next attack, against a German cruiser. The Navy Department objected to the size of the bombs, but the army, which was more than just a little tired of navy complaints, pointed out that in a real war the United States would use everything in its arsenal to defend the nation, including poison gas, if necessary. Then the navy announced that the objective was to note damage, not sink the ship. Detecting panic in the Navy Department, Mitchell attacked on July 18 in several waves. The navy tried to call off the last wave, but the Air Service had begun its final attack, sinking the ship. When asked why he did not stop the attack, Mitchell correctly pointed out that the ships were so far out to sea that when the planes were over target they had burned up almost half of their fuel. To stop a wave would have meant another day to complete the mission. As it was, some of the planes did not have enough fuel to reach Langley Field and had to land on the Virginia beaches.[11]

The great test was ahead with the huge German battleship *Ostfriesland,* which the navy claimed could not be sunk by aircraft. A number of naval officers, including Admiral Fullam, disagreed; it could be done, and it was. On July 19, Mitchell hosted General John J. Pershing, who had just become the chief of staff of the army, and the new Secretary of War, John Wingate Weeks, a no-nonsense businessman who was an 1881 graduate of the U.S. Naval Academy. Weeks had assumed the secretaryship on March 5 of that year. Although new to the job, he had made it known that he was unhappy with the inter-Air Service squabbling between Mitchell and Menoher and that he found Mitchell's private war with the navy to be unseemly. He was aware that admirals like Fullam were pushing the navy hard over aviation and were accusing the Navy Department of trying to silence dissent over airpower and submarines.[12] Pershing and Weeks toured the facility at Langley Field and inspected the 1st Provisional Air Brigade, which was tasked to carry out the bombing runs. At the end of the day the aviators of the brigade performed for Pershing and Weeks, showing how the various types of squadrons would act in concert to sink the great *Ostfriesland.*

On July 20 the bombing runs went well, with one of Mitchell's pilots hitting the ship with an 1,100-pound bomb. Other incoming planes were waived off by naval personnel, who boarded the ship to note the damage caused by a direct hit. Mitchell was furious that his planes had been ordered back to Langley Field with seven undropped bombs, but he could do little since a summer storm was approaching and the navy was taking a very long time to note all of the destruction wrought by the hit. Pershing, Weeks, Menoher, and others were on the USS *Henderson* watching this obviously successful attack, and they were joined by aviation pioneer Glenn Martin and naval Lieutenant Commander Zachary Landsdowne, who was Mitchell's special guest on the *Henderson.* Mitchell was determined that no navy rules or time delays would stop the next day's attack on the ship, which was already listing from the pounding she took the previous day. With Mitchell in the air, his pilots came to the target area at noon, and Captain W. R. Lawson commanded a flight with 2,000-pound bombs. Inspectors were again ready to board the *Ostfriesland,* but Billy Mitchell, sensing the moment, would have none of that. At twenty minutes after noon the great battleship took a direct hit, a mortal blow. The coup de grace was administered five minutes later, when a bomb exploded below waterline. Within ten minutes the powerful battleship began to roll, turning completely over. By one o'clock in the afternoon, with escaping air sounding like a giant sigh, the *Ostfriesland* went down to the bottom. There was silence on the *Henderson,* and some old sailors had tears in their eyes seeing this once proud, powerful, fear-inspiring enemy ship sink beneath the water. To rub salt into a deep navy wound at that moment, Mitchell, with pennants flying from his own aircraft, flew near the *Henderson* waving his wings in triumph.

When Mitchell's plane landed at Langley Field his young aviators, flush from a mission brilliantly accomplished, hailed Mitchell as their hero, carried him on their shoulders as a band played, and cracked open a number of now-prohibited bottles. This was for Billy Mitchell a highwater mark, the pinnacle of his career, surpassing even his massing of aircraft for the St. Mihiel operation; for Billy Mitchell on that noisy night of July 21, 1921, there was only back-slapping and many toasts. The next weeks were filled with congratulations for a test well conducted, for promises about airpower well kept. The British air attaché in Washington, Air Commodore L.E.O. Charlton, sent official notice of his enthusiasm for what he had witnessed from the *Henderson:* "I can not sufficiently admire the effort whereby so large a number of land machines were concentrated at such a distance to the coast. Hearty congratulations."[13] Lieutenant Colonel Aldo Guidoni, air attaché of the Italian Embassy, wrote, "Although the ability and efficiency of the United States aviators is well known in Italy, I did not expect such a precision in the results."[14] Admiral William A. Moffett, soon to be the navy's chief of the Bureau of Aeronatics, understood that Mitchell had won a public victory of monumental proportions against the navy in general and for army aviation in particular.[15] In August the widely read magazine *Aviation* carried a statement that Mitchell had made in April about upcoming bombing trials. The editors commented, "On reading it in the light of the recent bombing experiments with warships it is difficult to realize that his claims at the time have been characterized as those of a fanatic."[16]

Newspapers from the giant *New York Times* to Mitchell's hometown *Milwaukee Sentinel* to the *San Francisco Chronicle* praised the general's very obvious successes over the water off the Virginia beaches. Requests from every type of citizens' group for Mitchell to come and speak poured into Washington. Of course Mitchell was overjoyed at the attention, which far exceeded anything his senator-father had enjoyed during his time in public life, but there were unsolved issues for Mitchell. His relationship with Menoher had to be settled, and Billy Mitchell understood that his marriage with Caroline had deteriorated to the point of no return. While Pershing and Secretary Weeks were obviously impressed with the power of the 1st Provisional Air Brigade, they had to make some decisions about Mitchell. To make matters worse for Pershing and for Weeks, the Mitchell-Menoher squabble had become very public and very nasty. There was little question that Mitchell was now impossible to control, but Menoher's leadership as chief, Air Service, was not especially forceful. His brother-in-law and executive officer, Colonel William Pearson, was disliked by many in the Air Service because of his own overbearing personality. Mitchell was maneuvering in Congress to create a groundswell for making him chief in place of Menoher.[17]

In early June Menoher asked Secretary Weeks to relieve Mitchell and assign him to an air post away from Washington. The request, much to Weeks's disgust, found its way into the press, and Weeks told reporters that he would investigate the matter.[18] John Weeks now had to handle a potentially embarrassing situation, and he pointed out that both officers were men of courage in battle and both were soldiers who had commanded with success in the Great War. One of the great publicly discussed issues was the

employment of nonrated officers to hold high positions within the Air
Service, and Mitchell had already struck out at Menoher over the fact that
he knew nothing about the airplane or air combat.[19] The *New York Times* on
June 11, 1921, waded into the controversy with an editorial urging Weeks
to convince Menoher to withdraw his request that Mitchell be reassigned
duties outside of Washington. The tone of the editorial was that Mitchell
had commanded well in France, and although his ideas were not always
tactfully expressed he was worth too much to the army and to the country
to be sent away from the center of military power and decisionmaking in
Washington.[20]

On June 17 Weeks announced that he would issue his decision on the
Mitchell-Menoher controversy, and while he refused to release any details
to the press he hinted that neither man would be relieved of his post.[21] That
is exactly what happened on the next day, when Weeks announced that
Menoher would withdraw his request for Mitchell's relief and that Mitchell
was admonished to stop his public attacks on the chief of the Air Service
and to remember his duty as a subordinate army officer.[22] It was not a satisfactory settlement, but it was politically expedient to keep both men while
the Air Service was preparing for the bombing trials against German ships.
John J. Pershing did not like the inconclusive settlement, and he had in
mind a way to settle the controversy.

On August 29, 1921, Mitchell submitted his after action report on the
bombing exercises by the 1st Provisional Air Brigade. The report very
quickly found its way to *Aviation* magazine, and Mitchell's findings and
recommendations were reported verbatim from the document.[23] What
Mitchell said was nothing more than what he had stated publicly before
Congress and civic groups. Basically Billy Mitchell posited that airpower
was the way to defend the United States from an over-the-water attack, and
somewhat surprisingly Mitchell argued that "Navy control so far as it
affects coast defense should cease 200 miles from the coast."[24] The report
also hinted that lighter-than-air ships—dirigibles—should belong to the
army and not the navy, as had been specified by Congress. The problem
was not what Mitchell wrote to Menoher but how quickly it came into the
possession of civilian editors. Pershing decided that the time had come to
end the problems within the Air Service, and he decided to take the same
actions that he had in France in late 1917.

When Black Jack Pershing became chief of staff in July, he brought
Major General James Guthrie Harbord with him as his assistant. Pershing
had brought Harbord with him to France in 1917, and Harbord had func-
tioned well as the chief of staff of the AEF, had commanded the 2nd

Division in combat, and, at Pershing's request, had taken command of the Service of Supply. Harbord was a good soldier, a first-rate administrator, and totally loyal to his commander. When the Air Service of the AEF floundered in the late winter of 1917, Harbord and Pershing had agreed that Mason Patrick should be brought in as chief. This would be Pershing's solution again—bring Patrick back as chief of the Air Service. Menoher asked to be relieved as chief and sent back to command troops in the field where, as he told Weeks, he had wartime experience and success.[25] The waters immediately became muddied when reports reached the press that Mitchell had allegedly asked to be removed from his post as assistant to Menoher.[26] Now-Colonel Patrick received a telephone call from Harbord to report to General Pershing's office on September 21. It was Harbord who offered the command of the Air Service to Patrick, who was not at all sure he wanted to take charge at a time when there was disorganization and very heated antagonisms within the Air Service. His argument to Harbord was simple—he had been assigned to straighten out the mess once, in 1917–1918, and he was not sure that he wanted to try to control officers like Mitchell again.[27] The *New York Times* had already cited reports to the effect that if Mitchell were not made chief of the Air Service he would resign and leave the army.[28] Mitchell always had an offer of a well-paying job from Eddie Rickenbacker, who was involved with various automotive concerns promoting the use of the automobile, in his pocket, but a second-rank automobile company executive rarely found his name in the *New York Times* day after day. A feisty, intrepid airman could, however, be the darling of the press.

Mitchell sent a request for relief and reassignment to the adjutant general of the army on September 17, citing as his reasons "that the conditions with respect to the development of aviation, as they now exist in the War Department, make my presence in the executive organization a source of irritation rather than a means of progressive advancement."[29] The request irritated Weeks and Pershing, then in the process of replacing Menoher with a reluctant Patrick. Was this a move by Mitchell to have him made chief of the Air Service? The response was probably not what Mitchell had expected, for Harbord informed him that same day that Weeks wanted Mitchell to remain as the assistant while he was preparing for bombing runs against the obsolete battleship *Alabama*. Once the *Alabama* tests were completed, if Mitchell still wanted to be relieved he could resubmit his request and Weeks would "take such action as he deems to be in the best interest of the service."[30] It was fairly clear that Weeks, Pershing, and Harbord had had enough of the controversy, and that Mitchell would be an assistant, but not the chief. On September 22 the War Department announced that Mason Patrick, a colonel of engineers, would resume the post he had held for the AEF, and he would be elevated to the rank of major general.[31]

Patrick knew that his job would not be easy and that he would have to confront Mitchell, who was fresh from the *Ostfriesland* triumph and from General Menoher's resignation. Mitchell was his main concern, and he decided how he would handle him as early as possible. One of Mitchell's loudest criticisms of Menoher had been that he was not a flyer and so knew nothing about the air. Patrick had done well as chief of the AEF's Air Service, but he lacked the one thing that would bring him credibility—pilot's wings. At age fifty-eight Mason Patrick decided to take flying lessons at Bolling Field near Washington. There were those who were outraged at Patrick's appointment, however, and they made their feelings known. Editorials in newspapers blasted John Weeks, and congressmen grumbled despite the fact that the Senate approved Patrick's appointment without one dissenting vote.[32] The great war hero and air ace Eddie Rickenbacker, one of Mitchell's staunchest friends, digressed from a speech he was giving to the Automobile Dealers Association convention in Baltimore to rip into Weeks and the army. Speaking about Patrick, Rickenbacker said, "His appointment is as sensible as making General Pershing admiral of the Swiss Navy."[33]

Never one to shirk his duty, Patrick wanted a confrontation with Mitchell as quickly as possible, and he got it. Mitchell had drawn up a reorganization plan for the Air Service which, as Patrick read, essentially would place Mitchell in charge of the service. Patrick quickly disapproved the plan and told Mitchell in no uncertain terms that he was the chief and would be glad to have Mitchell's advice, his recommendations, and his plans. Mitchell then tried a grandstand play, stating that he should resign and requesting that they go to Harbord's office together. Determined to support Patrick, Harbord told Mitchell that if he chose to resign, his resignation would be accepted on the spot, whereupon Billy Mitchell quickly retreated and decided that he could indeed support Patrick.[34] Mason Patrick was a good, strict soldier, but he was not a vindictive man, and he firmly meant to bring order out of the confusion in the office of the chief. Patrick went back to the plan that Mitchell had presented and he had disapproved, and he then accepted a plan that called for three distinct sections within the Air Service: war plans, training, and a large umbrella section that would deal with questions of airways, meteorology, and communications. He left it to Mitchell to fill those positions, but he required full consultation prior to each assignment. Mitchell agreed to the arrangement and promised that all of their decisions would be recorded in memo form so that no misunderstanding might arise.[35]

Patrick, Harbord, Pershing, and Weeks were hopeful that finally the Air Service might be on even ground, with Patrick in full charge. Patrick had seen Mitchell function well in combat, and he felt that Mitchell could serve the nation well as long as his supervisors understood that he had a good and productive side in addition to a dark and rebellious side.[36] But nothing in their military relationship could have prepared Patrick, Pershing, or the army for what was to be a massive upheaval in Mitchell's personal life. After Billy Mitchell had moved out of his home and Caroline had taken the children back to New York, he appeared to friends to be acting even more erratic. Concerned over her husband's behavior, Caroline returned to Washington and sought out Senator James W. Wadsworth of New York. The relationship between the Mitchells and the Wadsworths developed because of a mutual love of horses and because Wadsworth had been impressed with Billy Mitchell's views on airpower. Mitchell's constant talk about a political career after the *Ostfriesland* sinking and the *Alabama* tests bothered Wadsworth, who did not see Billy Mitchell as a candidate for the U.S. Senate from Wisconsin. Caroline Mitchell met with Wadsworth to express concern over "her husband's mental and nervous condition."[37] Evidently Wadsworth, who liked Caroline more than Mitchell, advised her to see Black Jack Pershing about the situation. In late November 1921 she met with Harbord to seek assistance from him, from Pershing, and from Secretary Weeks. Pershing and Weeks had thought that problems with Mitchell had subsided, but now they were faced with another distasteful situation.

After Caroline had presented the situation to Harbord he decided something had to be done. He asked Senator Wadsworth his opinion, and Wadsworth concurred that Billy Mitchell's behavior was cause for concern.[38] Captain W. H. Frank delivered an order to Mitchell in his office telling him to report to Walter Reed Army Hospital for mental observation and possible treatment, "with a view to his being put into the psychopathic ward."[39] The order came from Harbord's office, but it could not have been issued without Pershing's agreement. Mitchell immediately called Harbord about the order. He was probably shocked to learn that Harbord had conferred with Senator Wadsworth and with Secretary Weeks. So seriously did Harbord take the situation that he went to John Week's personal residence, where the secretary was recuperating from an illness, and Weeks agreed with the order. On November 25, Mitchell sent two telegrams, one to his mother telling her about the situation and another to his sister Harriet asking her to come immediately to Washington to help him.[40] To complicate matters, Patrick had decided in early November that Mitchell's time would be best spent in a fact-finding mission to observe what European states were doing in aircraft development and in refining airpower doctrine. Patrick saw this tour as a way to get Billy Mitchell out of Washington and out of the United States for an extended period.

In what must have appeared to Mitchell's mother as deja vu, remembering her own husband's nasty divorce, the crisis now required a circling of the Mitchell wagons. Mitchell told his mother that he wanted "an absolute divorce in Wisconsin and try to obtain the two younger children Harriet and John. The important thing however is to get the divorce."[41] Billy Mitchell never explained why he was willing to fight for two children while he and Caroline had had three. In a few days Harriet Mitchell arrived in Washington. "Willie was ready to hit back. We had plans all made for him after leaving the army and I certainly hope he would get out," Harriet wrote to her ailing mother back in Milwaukee. On November 30, Mitchell met for half an hour with Secretary Weeks while Harriet waited in the outer office. Pershing was now involved in requiring that Mitchell agree to a battery of psychological tests, and Harriet was asked about both her brother's and Caroline's mental conditions. In the same letter to her mother, Harriet stated a very revealing fact when she wrote, "W [Mitchell] has hardly touched a drop since I have been here and he is so well. Yesterday he ate the best meal for luncheon I had seen him eat."[42] If there had not been drinking problems with Mitchell, Harriet, who doted on her older brother, would never have mentioned drinking to her mother.

As the tangled mess got worse Harriet continued to support her brother, blaming everything on Caroline and her Rochester and Washington lawyers. Caroline's complaints included extravagance, taking their daughter Harriet flying without permission, and "telling of things while intoxicated." The loyal sister believed that her brother was the victim of War Department intrigue and jealousy. Mitchell contemplated a lawsuit against his estranged wife for slander, but she had given her lawyers all of the family's canceled checks. She refused to pursue any claims of adultery, however, even though she had evidence of a possible occurrence in Baltimore. The situation had grown into a public scandal, and this displeased Pershing, who feared for the reputation of the army. As deliberations went on Harriet told her mother, "Well I guess our little general has put it over the old men again and they are afraid of him. But now Willie may have to resign and come home to live after this is all cleared up."[43] This was far from the truth, however, because Pershing and Weeks wanted to dispose of this messy situation, and as long as Pershing was chief of staff Mitchell would never become the chief of the Air Service. The best solution was to allow Patrick to send Mitchell on his European tour and allow the divorce to proceed through Wisconsin courts.

Mitchell met with Patrick in early December and learned that he would be sent to Europe. The Mitchell clan totally misunderstood the position of the army about the situation. They did not, as Harriet believed, "all like to be on the band wagon. Well they are afraid of Willie too."[44] Mitchell filed for divorce in Wisconsin prior to his departure for Europe, charging Caroline with cruel treatment, and she countered with a suit citing almost all of their marital problems. When the case finally came before the judge, he asked Mitchell if indeed he had not walked out of the family home on June 14, 1921, and Mitchell had to admit to this and to other problems. With desertion on the record and other testimony aired, Caroline was granted a divorce on Mitchell's home ground of Wisconsin. He was ordered to pay $150 per month for each of the older children—Elizabeth, age 17, and Harriet, age 13—and $100 per month for the child John, age 4. Mitchell was able to counter alimony claimed by Caroline, citing her personal property valued at a quarter million dollars, but time would show she was not through with personal alimony.[45] The final decree went into effect on September 22, 1922.

The divorce is very revealing about Mitchell, his family, and their view of him as a major power within the army at that time. According to the family, Billy was a victim of jealous old officers in the War Department, and despite evidence to the contrary, the actions of Caroline were not based on their real relationship. They believed that Caroline Stoddard Mitchell had become a tool for those who wished to destroy Billy Mitchell's career. A thread of fear of Mitchell ran through the whole tapestry of the messy divorce, and certainly neither Pershing, Harbord, nor Weeks wanted Mitchell marching up to Capitol Hill to demand a Senatorial investigation, as he had threatened. But with Wadsworth and others now disgusted with Mitchell's grandstanding and with his attacks on the military in a time of limited budgets, the possibility of an investigation benefiting Mitchell was small indeed. General of the Armies John J. Pershing prized teamwork and loyalty, both personal and to the institution of the army. In this case that worked to Mitchell's advantage, because Patrick's solution was simply to get Mitchell out of the United States and out of the glare of the public press.

There were dangers in sending Mitchell abroad to confer with other aviation experts and in his attending a conference on the limitation of armaments. Mitchell was at the peak of his notoriety after the *Ostfriesland* success, and a Mitchell basking in the adulation of fellow air officers in

Europe, with wines flowing freely, could spell disaster. Patrick sent him a memorandum that spelled out in very clear terms what Mitchell was not to do while in Europe:

> It is especially desired that to invite your attention to the necessity for being very discreet. . . . [I]t will be most unwise to indulge in any discussion or criticism of the Air Service policies of the United States or to make any statement concerning what the future policy will or should be. While satisfied that you already grasp the facts recited above, it is thought well to put on record the fact that your mission is primarily and exclusively the gathering of information and not the discussion of Air Service or aeronautical policies.[46]

This was not the memorandum of an old man who was afraid of Mitchell. It was, in fact, a warning from Patrick, Pershing, and Weeks that any further violation of military protocol would not be tolerated. The European trip, which some believed was a sort of rest cure in lieu of the psychopathic ward at Walter Reed Army Hospital, was actually a trial for Mitchell as an officer in the Air Service. There was no doubt that Mitchell, with his command of French, his many ribbons and decorations, his natural command of social situations, his debonair walking stick, his reputation as a fine war fighter in the air, could be charming and sophisticated, the type of American to whom Europeans could be attracted. But would that dark side—the unwillingness to remain in a harness prescribed by the upper echelons of army command—take over and cause Mitchell to overstep Patrick's instructions?

Mitchell took First Lieutenant Clayton L. Bissell and aeronautical engineer Alfred Verville with him. Both men were close to Mitchell, Bissell having hit the *Ostfriesland* on July 20. The itinerary was a busy one, with a full tour of Italy, including visits in Rome, Milan, and Turin; and a major stop in Germany visiting sites in Berlin, Coblenz, and Friedrichshafen. Of course there were meetings in Paris and in Britain. The Atlantic crossing in January was uneventful, as was Mitchell's first conference in Paris, but in Italy Mitchell was greeted as a great hero. The Roman newspaper *Il Piccolo* on January 17, 1922, carried a full picture of Mitchell on the front page with the headline "Rome Salutes the Representative of the Valiant Army of the United States." The accompanying story stated that Mitchell was greeted by a large and enthusiastic crowd led by the prime minister of Italy and members of the Italian Air Staff.[47] *Il Messaggero* of Rome on the same day was equally effusive, calling Mitchell the "vice capo" of the U.S. Air Service.[48] This was heady stuff for Mitchell, whose ego fed off of adulation and press coverage. Consequently, Mitchell was very receptive to Italian ideas of airpower doctrine, and much of what he said and wrote after the 1922 trip would reflect this spectacular outpouring by the Italians.

Billy Mitchell was at his best, pleasing his Italian hosts with his praise for the "incomparable beauty" of Rome.[49] But it was his visit to the offices of the Caproni Company in Milan on January 28 that had the greatest impact.[50] The Caproni bombers were considered some of the best in Europe at that time, and it was there that Mitchell heard again the concepts of strategic bombing, which he had first encountered over lunches and in meetings with General Hugh Trenchard in England in 1919. It was obvious to the Italian press that the developers of the Caproni aircraft saw in Mitchell a fellow aviator who could understand their goals and capabilities. An air attack on the United States by an enemy launching planes from great distances now became a vital part of Mitchell's arguments for a separate air force. Before he left Italy Mitchell gave a lengthy interview to *La Gazzetta dell'Aviazione* that was published in early February. Mitchell was very careful in this interview to follow Patrick's clear admonitions.

Mitchell responded to the interviewer by praising the Italians for their work on dirigibles, especially the airship *Roma*. He restated his belief that airships were vitally important in the development of air defense and offensive operations. Choosing his words carefully, Mitchell stated that American aviation in the area was slow and that time would be needed to convince the government to proceed in the development of American airships. When asked, Mitchell avoided any direct criticism of Josephus Daniels and the Navy Department, but did recount the sinking of the ships as a triumph for airpower. This was pretty tame stuff for Billy Mitchell when given a chance to shine before an appreciative audience; he did have lavish praise for Gianni Caproni and hinted that Caproni and his ideas were the way of the future, but that was as far as Mitchell could go at the time in dealing with the concepts of strategic bombings.[51]

The remainder of the lengthy mission was unmarked by any explosion, gaff, or unguarded statement by Mitchell. His orders, issued by the War Department, sent Mitchell to Friedrichshafen, Germany, to observe the operation of the Zeppelin Company, which was considered to be one of the best in the world. The navy had already expressed an interest in obtaining a zeppelin, and Admiral William A. Moffett was ever on guard against Mitchell's desire for a unified air service that would include lighter-than-air ships rather than placing them under navy control.[52] There is no doubt, given Mitchell's subdued statements in Italy, that he viewed the zeppelin craft as vital to American airpower. The situation between Moffett and

Mitchell over this issue would continue to fester until it became a serious cancer for both the army and the navy. But during this trip to Europe, Mitchell would give no offense to Moffett, the navy, Secretary Weeks or anyone else simply because he was enjoying being the center of attention. American newspapers such as the *New York Times* and the *Chicago Tribune* carried stories about Mitchell's mission to Europe with such items as his being the guest of honor at various Italian air shows and meets.[53]

Meanwhile Patrick, Harbord, and Pershing had come to the conclusion that the best way to harness this energetic and highly intelligent officer was to keep him on inspection tours and overseas missions. When Mitchell returned to Washington he gave every indication of being a good subordinate, free of controversy for the time being. With his final divorce hearings pending in Milwaukee, Mitchell carried on a bachelor officer's life, devoting a good deal of his time to horses and to social events, where he cut a handsome picture of a man in the full vigor of life. In August 1922 Patrick sent Mitchell as the Air Service representative to an air meet at Selfridge Field near Detroit, Michigan, and there he met Elizabeth Trumbull Miller, the daughter of a prominent Detroit lawyer. Fifteen years younger than Mitchell, Elizabeth was a striking brunette with a love of horses and a fascination with aircraft. Her mother was Lucy Trumbull Robinson (of Hartford, Connecticut), and her father, Sidney T. Miller, was a Harvard Law School graduate who had returned home to begin a successful career as a corporate lawyer. Miller was a founder of the Detroit Symphony and a past president of the Detroit Bar Association. Elizabeth and Billy were immediately attracted to each other and soon were contemplating marriage. Her parents insisted that all of these plans had to remain secret until Mitchell's divorce became final and a suitable period had passed. As Billy Mitchell would find out, Elizabeth was captivated by the Washington social scene and by raising fine horses in the right surroundings.

In September Mitchell took a short leave and participated in the fall fox hunt at the Harford Hunt Club north of Baltimore. In typical Mitchell fashion, he was attired in the best of British riding habits and brought four of his finest horses. He could not keep Elizabeth far from his mind, and he planned to spend some of his leave in Detroit in order to see her. Mitchell wrote to his friend and disciple Captain Carl Spatz (he later spelled his name "Spaatz" and was the first chief of staff of the U.S. Air Force) if he would ask a good downtown hotel "to reserve a small room and bath in a quiet location for me." He also asked Spatz to reserve a car for him. Regarding any question as to why the hush-hush about the week-long stay in Detroit, Mitchell said that he did not "want a lot of noise made about my arrival."[54] With a new love in his life but without a final divorce decree, it was evident to Mitchell that his probable future in-laws would not want much notoriety either. The fall of 1922 was a good one for Billy Mitchell. On September 22 his divorce from Caroline was made final by a court in Milwaukee. She had won custody of the children and child support (which he was often slow in paying and sometimes did not pay at all), but he was now free to pursue his blossoming relationship with the pretty Detroit socialite. Prior to his quiet September visit to Detroit, Mitchell traveled to Maine for trout fishing, telling Carl Spatz, "I have been packing through the country with one guide. I have had great trout fishing and I feel like a fighting cock."[55] In late mid-November Mitchell took five horses to the National Horse Show in New York City where he planned to show all of them, especially his favorite gelding, Sir Dixon II, a handsome sorrel.[56] Professionally things had gone well with his chief, and in early October Patrick appointed him to be on the Air Service Technical Committee to standardize the equipment of the Air Service.[57]

During the Christmas season Mitchell looked forward to a visit from his mother, who had not been well. It must have been a strange Christmas for Mitchell, who now occupied a small home in northwest Washington. During the visit Harriet Mitchell had appeared to be tired and sometimes vague, and on December 28 she took a turn for the worse; before Mitchell could summon an ambulance, she was dead. This was a great blow for Mitchell, who had always been able to confide in "Dear Mummy," a term of endearment he used up to her death. She had been the one who had, at Meadowmere back in Milwaukee, introduced him to fine horses. When the young Willie had fallen off a horse, she simply told him to dust himself off and get back on. Harriet Mitchell had been Mitchell's correspondent while he was a student at Racine College and then a callow lieutenant in Cuba and the Philippines, and she had done much to compensate for the distant father who never really approved of his son's coat of blue with officer's shoulder straps. Perhaps unwisely, she had supported her son financially when his way of life tended to the extravagant. (Indeed, when he had asked for a baked possum as a treat while at college, she had made sure it was done and sent properly.) Now that she was gone, Billy Mitchell would rely on his two doting sisters—Harriet and the outspoken and adventurous Ruth—who would always respond to the older brother they saw as their "little general" bravely jousting with the old mossbacks of the War Department. And perhaps Elizabeth Miller of Detroit could be the ballast that would keep Mitchell steady and out of trouble. That, only time could tell.

NOTES

1. Mason Patrick, *The United States in the Air* (Garden City, NJ: Doubleday, Doran and Co., 1928), 76.

2. "General Mitchell on Aviation," *Aerial Age Weekly* 12, 23 (14 February 1921), 579.

3. Press Release by Fullam, 30 March 1921, in the William F. Fullam Papers, Library of Congress, Washington, DC, Carton 6. (Hereafter cited as the Fullam Papers.)

4. Donald Wilhelm, "Two Dimensions Vs. Three," *The Independent* (7 May 1921), 477, 500.

5. William Mitchell, "Has the Airplane Made the Battleship Obsolete?" *World's Work* (April 1921), 550–555.

6. Mitchell to Bane, 5 March 1921, in the William Mitchell Papers, Library of Congress, Washington, DC, Carton 7. (Hereafter cited as the Mitchell Papers.)

7. Fisher to Mitchell, 11 June 1921, Mitchell Papers, Carton 9.

8. Harriet Mitchell to Her Mother, 5 December 1921, Mitchell Papers, Carton 19B.

9. Clipping Files, in the George Hardie Collection, Golda Meir Memorial Library Archives, University of Wisconsin, Milwaukee, Carton 10, Folder 8. (Hereafter cited as the Hardie Collection.)

10. Mitchell to Mother, 3 August 1921, Mitchell Papers, Carton 19B.

11. William Mitchell, *Winged Defense* (New York: G. P. Putnam's Sons, 1925), 63–64.

12. Fullam to Secretary of the Navy, 26 May 1921, Fullam Papers, Carton 6.

13. *New York Times,* 30 July 1921, 3.

14. A. Guidoni, "A New System of Bombing Tests," *Aviation* (3 October 1921), 307.

15. William F. Trimble, *Admiral William A. Moffett: Architect of Naval Aviation* (Washington: Smithsonian Institution Press, 1994), 88–89.

16. "What General Mitchell Claimed," *Aviation* (1 August 1921), 133–134.

17. W. H. Frank to Mason Patrick, 28 January 1928, in the Benjamin Foulois Papers, Library of Congress, Washington, DC, Carton 5. (Hereafter cited as the Foulois Papers.)

18. *New York Times,* 10 June 1921, 2.

19. Ibid., 11 June 1921, 3.

20. Ibid., 12.

21. Ibid., 17 June 1921, 3.

22. Ibid., 18 June 1921, 2.

23. "General Mitchell Attacks Bomb Test Findings," *Aviation* (3 October 1921), 367.

24. Memorandum from Mitchell to Menoher, 29 August 1921, Mitchell Papers, Carton 40.

25. *New York Times,* 17 September 1921, 3.

26. Ibid., 21 September 1921, 19.

27. Patrick, *United States in the Air,* 82–83.

28. *New York Times,* 22 September 1921, 4.

29. Mitchell to Adjutant General, 17 September 1921, Mitchell Papers, Carton 9.

30. Harbord to Mitchell, 17 September 1921, Mitchell Papers, Carton 9.

31. *New York Times,* 22 September 1921, 4.

32. Ibid., 1 October 1921, 14.

33. *Baltimore Evening Sun,* 27 September 1921, 1, in the Mitchell Papers, Carton 9.

34. Patrick, *United States in the Air,* 87–89.

35. Memorandum from Mitchell to Patrick, 18 October 1921, Mitchell Papers, Carton 11.

36. Patrick, *United States in the Air,* 86.

37. Martin L. Fausold, *James W. Wadsworth, Jr: The Gentleman from New York* (Syracuse: Syracuse University Press, 1975), 150.

38. Mitchell to Mother, 25 November 1921, Mitchell Papers, Carton 19B.

39. Frank to Patrick, 28 January 1928, Foulois Papers, Carton 5.

40. Mitchell to Mother, 25 November 1921, Mitchell Papers, Carton 19B.

41. Ibid.

42. Harriet Mitchell to Mother, 5 December 1921, Mitchell Papers, Carton 19B.

43. Ibid.

44. Ibid. This letter is different from the one cited in note 43.

45. Clipping from a Detroit Newspaper, ca. 1923, in the Hardie Collection, Carton 10, Folder 8.

46. Patrick to Mitchell, 7 December 1921, Mitchell Papers, Carton 11.

47. *Il Piccolo,* 17 January 1922, Clippings in Mitchell Papers, Carton 11. (Translations from the Italian are the author's.)

48. *Il Messaggero,* 17 January 1922, Clippings in Mitchell Papers, Carton 11.

49. Ibid.,18 January 1922, Clippings in Mitchell Papers, Carton 11.

50. *La Gazzetta della Sport,* 28 January 1922, Clippings in Mitchell Papers, Carton 11.

51. "Parlando col Generale William Mitchell," *La Gazzetta dell'Aviazione,* 6 February 1922, Clippings in Mitchell Papers, Carton 11.

52. Trimble, *Moffett,* 128–129.

53. *Chicago Tribune,* 18 January 1922, Mitchell Papers, Carton 11.

54. Mitchell to Spatz, ca. 1 September 1922, in the Carl Spaatz Papers, Library of Congress, Washington, DC, Carton 2. (Hereafter cited as the Spaatz Papers.)

55. Mitchell to Spatz, 4 September 1922, Spaatz Papers, Carton 2.

56. Mitchell to National Horse Show Association, 24 October 1922, Mitchell Papers, Carton 9. (This also includes his detailed entry form for the five horses.)

57. Personnel Orders No. 209, 11 October 1922, Spaatz Papers, Carton 2.

Colonel Billy Mitchell, 1925
(Courtesy of the Library of Congress)

John Lendrum Mitchell and William, c. 1881
*(Courtesy of the George Hardie Collection,
Golda Meir Memorial Library Archives, University of Wisconsin–Milwaukee)*

Mitchell and his daughter Harriet, c. 1910
(Courtesy of the George Hardie Collection,
Golda Meir Memorial Library Archives, University of Wisconsin–Milwaukee)

Mitchell and his staff, 1918 (left to right): Thomas D. Milling, Mitchell, and Paul Armengaud *(Courtesy of the Library of Congress)*

Battleship under air bombardment, 1921
(Courtesy of the Library of Congress)

Mitchell and his personal aircraft, 1924
(Courtesy of the Library of Congress)

Mason Patrick, 1925
(Courtesy of the Library of Congress)

Mitchell (far right in uniform) and his defense team at the 1925 court-martial (left to right): Colonel Herbert A. White, William H. Webb, and Frank Reid (*Courtesy of the Library of Congress*)

Hugh A. Drum, 1918
(Courtesy of the U.S. Army Military History Institute)

Dennis Nolan, 1918
(Courtesy of the Library of Congress)

Betty and Billy Mitchell, c. 1928
(Courtesy of the Library of Congress)

EIGHT

PACIFIC TOUR

T HERE WAS A NEW WIND BLOWING IN THE CHIEF OF THE Air Service's office with Major General Mason Patrick, who was doing exactly what Black Jack Pershing wanted him to do: make the Air Service a viable, disciplined military unit capable of playing a role in the defense of the United States. Pershing never lost his cherished belief that ultimate victory depended on the infantry, and to him the fount of all military knowledge still rested at the Infantry School at Fort Benning, Georgia, where the lessons learned from the Great War were studied and taught. In 1919, shortly after the troops returned from France, Mason Patrick had sent a long letter to his former AEF commander-in-chief in which he stated in no uncertain terms:

> It is understood that there has grown up in the Air Service itself a strong feeling on the part of some of the officers that their salvation lies in such a separate department. I cannot help thinking that most of them are influenced more largely by their own personal desires and ambitions than by regard for the best interests of the United States. . . . A common-sense view of the matter is that a normal, even though possibly somewhat slow development, and an absolute subordination of personal feelings and ambitions are necessary if this matter is to be handled and decided aright.[1]

At that time, Patrick did not believe in the idea of a separate or unified air arm. Now, as chief of the Air Service, faced with peacetime technological changes and an ever-expanding role for the air in military and civilian matters, Patrick was slowly changing his position and was becoming the spokesman for the expansion of army air.

With Mitchell in check for a while, Patrick could focus on what he felt was one of the most critical problems for all of the army—low morale in

151

the face of stagnant low pay, reductions in rank, and outdated equipment.[2] In October 1922 Patrick encouraged his subordinate commanders to make certain there was time on a daily basis for physical exercise and competitive games to encourage a sense of esprit de corps, "a sort of family feeling, which will go far to make them more contented, their lives better and happier."[3] He recalled that after the guns fell silent on the Western Front the troops remaining in Europe not only trained but also engaged extensively in competitive sports—baseball, football, and boxing. It worked well there, and it was worth trying again. Mason Patrick also knew that one of the great pressures on a professional soldier was a disgruntled wife and a dispirited family. This was particularly critical for the army during the Roaring Twenties, when prosperity was everywhere but in the armed forces, or at least many professional officers believed that to be the case. He also knew that many wives feared for their husbands' safety while in the air. In an extraordinary move Patrick issued a letter indicating that "I am considering, therefore, granting permission to those women members of an officer's immediate family who desire to take advantage of it be taken up in aircraft once to twice."[4] Too many good aviators and ground mechanics were leaving the service for higher paying jobs in private industry. Mitchell always had a standing job offer from his old friend and solid supporter Eddie Rickenbacker, and Rickenbacker continued to urge him to leave the army for a lucrative position in the automotive industry, which was flourishing in the 1920s.

Patrick's approach to low morale and starvation military budgets was quite different from Mitchell's. In letters to fellow officers like Thurman Bane at McCook Field or Carl Spatz at Selfridge Field, Mitchell was almost cavalier in his approach to their problems. When Mitchell was in the middle of the *Ostfriesland* and *Alabama* operations he telegraphed Bane to demand some information, asking in an irritated tone why Bane had not promptly responded to an earlier request. Bane replied that for over six weeks no one could send a telegram from McCook Field simply because there was no money available to pay the telegraph company.[5] Mitchell's response was basically that Bane, who had his rank and pay recently reduced from colonel to major, should personally fund the telegrams. Patrick's approach, in contrast, was to focus on what could be done to aid soldier and family morale while continuing to ask Congress to raise the Air Service budget.

Mitchell spent a good part of the late winter and early spring of 1923 engaged in inspection trips to various air fields. His detailed reports reflected what Patrick already knew—that there were morale and equipment problems. When not inspecting, Billy Mitchell spent a good deal of time being the horseman, the hunter, and the fisherman. He was also engaged in two projects: writing a book, and convincing the Millers to announce the engagement of their daughter. The book would be a rehash of what Mitchell had been saying for some years about a united air service, and a description of the *Ostfriesland* sinking and the *Alabama* tests. He had not given up on his ideas and believed that when the book was published it would win him a national following that might propel him into a newly created cabinet position for air matters. Throughout the end of 1922 and into 1923, however, it was Patrick and the navy who were winning headlines. A conflict festered under the surface, and it threatened to burst to the top at any time, given the right circumstances.

In early June Mitchell decided to enter the Detroit Horse Show, which was scheduled to take place at Grosse Pointe, Michigan, not far from the Miller home. It was typical Mitchell extravagance in that he shipped six horses and his "head groom" to Detroit, and he called on Major Carl Spatz, who was now commanding the Selfridge Air Field, to supply him with several soldiers to work with the groom and transport the horses to the stables. Mitchell also asked for more soldier time to make repairs on the stables and for forage to be brought with them for his horses; and he even asked Spatz to see if there was a government veterinarian nearby to tend to his animals during the show. He had also asked Eddie Rickenbacker to help him with the loan of an automobile, which his old friend agreed to do.[6] What Spatz or his soldiers thought of Mitchell's requests was not recorded, but the demanding Mitchell was asking for government forage, soldiers, and a veterinarian as if it were his due.

Poor Spatz was put in a bad position by Mitchell over this unofficial trip. Most of his soldiers had gone to support National Guard training at Camp Custer, Michigan, but he nonetheless found three soldiers to repair stables and help Mitchell's groom. As far as veterinary service was concerned, the two stationed at Fort Wayne had also gone to Camp Custer, but if Mitchell needed one he could be flown from Camp Custer to Detroit. The quartermaster at Selfridge Field took umbrage at the request for free fodder and told Spatz that if Mitchell needed fodder he could obtain it at army contract prices. Major Spatz informed Elizabeth Miller—or Betty, as Spatz called her—about Mitchell's plans, and he made arrangements concerning the automobile that Eddie Rickenbacker had agreed to let Mitchell use during the horse show.[7]

The Millers finally agreed to a proper time to announce officially the engagement of their daughter to Mitchell. They wanted to wait until a year had passed after the messy divorce went into effect, for the sake of society and propriety. But they agreed with Mitchell and Elizabeth that the engagement should be announced in mid-August, with a wedding to follow in early October 1923.[8] On August 14, the society section of the *Detroit Free Press* ran the announcement and told how the two met at the air races at Selfridge Field. The newspaper left out the fact that Mitchell's divorce had not been final at that point, nor did it see fit to discuss the messy details surrounding it. The fact that Mitchell had three children—one of them still very young—was not mentioned either.

Patrick saw the upcoming nuptials as a great opportunity to get Mitchell out of Washington on a lengthy honeymoon that could be coupled with an inspection of air facilities in the Pacific. It had been some time since anyone in the Air Service office had actually assessed airpower in that area, especially in Hawaii and in the Philippine Islands, so why not Mitchell? For Mitchell the idea was a godsend because it would cut down the costs of the trip. He would be under military orders and would only have to pay for Elizabeth's fare. They would stay in government-provided quarters, as befitted a general officer inspecting facilities.

Everyone with any sense in the War Department knew that the question of airpower was boiling beneath the surface and could explode at any time. There was a bitter fight raging in the navy over airpower, and since the *Ostfriesland* coup by Mitchell, lines had hardened. A number of aviation-minded admirals, including William F. Fullam and William S. Sims, constantly pushed for more attention to be paid to naval aviation and to the concept of the aircraft carrier. Admiral Moffett, chief of the navy's Bureau of Aeronautics, was doing as much for navy air as was Patrick for the Army Air Service, but both men were frustrated by shrinking budgets and public apathy about air defense. Mitchell had complicated the picture with his exaggerated and very generalized claims after the sinking of the *Ostfriesland*. When Admiral Fullam wrote to Assistant Secretary of the Navy Theodore Roosevelt Jr. about the future of air and aircraft carriers, he received a terse reply: "I object most strenuously to those individuals who devote no attention to the development of aviation but I object equally strenuously to the individuals who would abandon the battleship in favor of aviation. . . . There is nothing which indicates the battleship will not be, at least for the immediate future, the deciding factor."[9] His father, the late former president Theodore Roosevelt, who had held the same post and who helped build the powerful battleship navy, would have been proud of his son in his defense of the great surface ships.

Every part of the military establishment of the United States was fighting for every dime. Patrick was doing a good job of representing the Air Service's interest in a way that Mitchell never could, but he too was moving toward the conclusion that at some point the air arm of the nation would have to be united to avoid duplication, squabbling over spheres of influence, and wastage of precious money. This Patrick was a different commander from the man who wrote to Pershing in 1919 doubting the validity of a unified air service. He was not given to broad generalities as was Mitchell, but he could issue as many dire warnings about the future of air as his maverick assistant. In January 1923, in Boston, Patrick warned that in the very near future a "pilotless plane" could indeed bomb Boston or New York. He told his audience that advances in technology had reached a point where an "automatic air service" could launch without pilots, and, with modern communications systems, strike a target a thousand miles away.[10] In April, Patrick proclaimed that the "New Age of Transportation has come," and that air passenger service by airplanes and lighter-than-air ships would soon surpass railroads in speed and comfort.[11]

By June Patrick could sport the wings of a military aviator, earned at age sixty. Taught by the experienced aviator Major Herbert A. Dargue at Bolling Field, Patrick admitted he would never be an expert pilot, but those wings and his desire gave him credibility.[12] In July 1923 Patrick went to Capitol Hill to present his case for a yearly outlay of $25 million for the next two fiscal years. The need for new equipment and aircraft was the centerpiece of his presentation, but he pointed out that Air Mail service, the establishment of set airways for military and commercial aviation, spraying for insects, and scouting for forest fires also had a part in the Army Air Service request.[13] So compelling were Patrick's arguments that the *New York Times* carried an editorial supporting the request for increased congressional funding.[14]

It was a critical time to have a moderate but forceful voice for army air, because the navy was pushing for increased funding as well. Within the navy the aviation versus battleship controversy was raging. Admiral William S. Sims, head of the Naval War College at Newport, Rhode Island, saw the fight shaping up between the two sides. He wrote to retired Admiral William Fullam, the staunch airpower advocate, "The difficulty about the battleship business is that officers are actually afraid to express their opinions for fear of the doddering admirals on the Selection Board. Several of them actually told me this. Those who command battleships or who are to command battleships are afraid that any expression of opinion against them will hurt them."[15] It was also obvious that any time Mitchell attacked the battleship navy or crowed too loudly over the *Ostfriesland* or *Alabama* tests, the battleship supporters dug their heels in more. Patrick understood the need to fight forcefully for the air, but not to a point of starting a bloody army-navy war.

Meanwhile, Billy Mitchell was making typically elaborate plans for his wedding and subsequent tour of the Pacific. In typical Mitchell fashion he announced that after the inspection he might get in some tiger hunting in India, and Adjutant General Robert "Corky" Davis made certain that Mitchell could do that, but only at his own expense. To protect Mitchell and his wife in that turbulent area of the world, Davis did give Mitchell permission for an official visit to the British air base at Singapore and imperial military installations in India. Mitchell was ordered, however, to proceed to Japan after his inspections and his trip to India "to investigate Japanese Air activities and to China [to investigate] _ . . . the progress made along aeronautical lines in the country." However, the usually-efficient Davis failed to inform the State Department of these plans so that its staff could coordinate a very sensitive trip with the Tokyo government. Orders were prepared for the newly wed Mitchells to depart by military transport from San Francisco on or about October 23 for the Hawaiian Islands, the first stage of Billy Mitchell's inspection tour.[16]

At 4 P.M. on October 11, 1923, Billy Mitchell wed Elizabeth Trumbull Miller at the Grosse Pointe Protestant Church in Michigan. Billy Mitchell was a handsome, debonair figure in his full dress military uniform. The society editor of one of Detroit's newspapers described him as "nervous but genial," and described Elizabeth as beautiful, as indeed she must have been. The writer obligingly omitted any reference to the recent divorce. She did not have to, because another Detroit paper earlier had ungraciously outlined the messy proceedings. Washington Becker, Mitchell's uncle from Milwaukee, came to Grosse Pointe to drive him to the church, and after the ceremony there was a lavish reception, as befitted the social status, not to say wealth, of Mitchell's new father-in-law.[17]

If Major General Mason Patrick hoped to keep the lid on the Billy Mitchell pressure cooker, he was in error. Major General Charles Pelot Summerall was as good a soldier as the United States ever produced, and he was in command of the Hawaiian Military Department. A West Point graduate with a fine record prior to 1917, he commanded the 1st Infantry Division, "Pershing's Pets," during World War I. No one but the staunchest of Pershing disciples and a fine fighting soldier could have been given that honor. By the armistice of November 1918 Summerall was a corps commander, and in February 1919 he received his permanent star as a brigadier general, while men like Mason Patrick reverted to the rank of colonel. Charles Summerall was touchy, and intensely proud of any command he had. Usually they were the best in the army, simply because Charles Pelot Summerall made them that way. This was not a man to cross, but Billy Mitchell did exactly that. No one else would ever dare question

Summerall's preparations for war; Mitchell did, and it would cost him dearly.

No soldier ever likes a staff visit and inspection by higher headquarters. It disrupts schedules and training, and time must be taken to secure quarters and arrange entertainment, especially among general officers and most especially for a general officer with an attractive new wife. Summerall would provide, as was the custom of the service, everything that Mitchell needed, and indicated to him that his office was open to discuss Mitchell's findings and recommendations.

Summerall and Mitchell came to the inspection from two very different perspectives. Summerall was concerned about the current state of military defense on the Hawaiian Islands, whereas Mitchell was specifically concerned about the status of the Air Service units in Hawaii and about a possible enemy attack in the future. He proposed that an air attack launched from carriers or from enemy-held air fields on, for example, Midway Island could devastate military installations in Hawaii. Of course Mitchell had shown just what well-trained pilots could do to stationary battleships, and he easily imagined bombers and torpedo planes wreaking havoc on Pearl Harbor. Many Americans later would remember Mitchell's predictions, but Summerall had to deal with the immediate concern of maintaining the facilities on his small budget and with a dwindling supply of soldiers, since army recruiting in the prosperous 1920s was at a dangerous low.

Mitchell had a tendency to rely heavily on reports from officers of the Air Service, but he did not see fit to surface Air Service concerns with Summerall or his staff. Colonel George E. Lovell Jr., commanding officer of the Fifth Composite Group on Ford's Island, gave Mitchell a dismal report about conditions. "The general supply system for Air Service units in the Hawaiian Islands seems to be very poor," Lovell wrote, and he added, "We are approximately eighteen months behind in requisitions and in a great many cases when equipment arrives it is either obsolete or badly damaged on account of poor packing."[18] Lovell also hit a hard blow at Summerall's logistic officers when he stated that the commanding general had made an error in creating an Air Depot that was not at Luke Field and therefore was not especially responsive to the needs of the principle Air Service airfield in Hawaii. Lovell's report to Mitchell also noted that though Lovell had requested a boat for rescue services in case a plane went down in Pearl Harbor, Summerall had not answered three separate requests for such a craft.[19] All in all it was a dismal report that cast great doubt on Summerall's G-4, or staff logistics officer, and on Summerall's commitment to the Air Service. Mitchell incorporated this and much more in his

own report, which he gave to Summerall upon his departure for the
Philippine Islands in early December 1923. Mitchell had not discussed his
findings and concerns with Summerall in an exit interview, which was the
accepted method of ending an inspection tour.

As far as Summerall was concerned, the quality of training and maintenance in the Hawaiian installations was progressing. As required by
instructions from Washington, Summerall signed a monthly activity report
that detailed all Air Service activities during the preceding month. For the
month of October Summerall noted the arrival of Billy Mitchell and the
thirty-one planes from Luke and Wheeler Fields that greeted him.
According to Summerall's monthly activity report, Mitchell expressed his
pleasure at seeing so many aircraft flying complex tactical maneuvers.[20]
The activity report for December 1923 indicated a large number of air exercises flown, and Summerall ended it by calling attention to the critical need
for more parachutes for aircraft and for more qualified air observers to be
sent the command. In that report, Summerall stated that Mitchell
had telegraphed him upon his departure, "Pursuit and bombardment exe-
cuted splendid maneuvers yesterday. Many thanks to yourself and Mrs.
Summerall."[21] There was apparently no reason for General Summerall to
believe that anything was found wanting in Mitchell's inspection of the
command, but as he read Mitchell's report he realized that all was not as it
had seemed to be when Mitchell was in Hawaii. The worst part of the stun-
ning report was that Mitchell, a brigadier general, had not expressed himself in person to a superior officer who was charged with command of the
Hawaiian Islands' military defenses.

Summerall was furious with the report and immediately registered his
displeasure with Mason Patrick, who was no doubt irritated to see the hor-
net's nest that Mitchell's high-handed methods had stirred up:

> The duration of his [Mitchell's] visit was necessarily limited, and he
> expressed his regret at being unable to go more thoroughly into the vari-
> ous details and to confer more extensively with the various commanders
> and staff concerned, other than those of the Air Service. This unavoidable
> necessity is regrettable, since I feel that, on this account, General Mitchell
> did not receive a correct and comprehensive impression in all cases. . . .
>
> The report deals at some length with the actual condition of the Air
> Service here and emphasizes its inadequacy. Apparently, little or no cog-
> nizance is taken of the fact that the peace garrison of the Air Service is
> less than one third of the war garrison. The peace garrison, in reality, is
> only a nucleus of the war garrison.
>
> The report, in considering the war plans of the Air Service, ignores
> entirely the Naval Air Service. . . .
>
> It is believed that superficial impressions and academic discussions

may result in conclusions that are unfair to the command, whose officers and soldiers are laboring whole-heartedly to improve their efficiency and to fulfill their mission.[22]

Summerall's rejoinder to Mitchell's report ran seven single-spaced pages, and it was page after page of great anger. He was particularly irritated with Mitchell's harping on a unified air arm for the Hawaiian Islands, and he was livid when he addressed Mitchell's charge that "Air Service funds have been improperly diverted. Such a charge is unwarranted." That touched on Summerall's pride in running an exemplary command, and he was quite rightly infuriated. As for Mitchell's speculations as to what a future enemy might do, Summerall told Patrick that because no one knew what might happen in the future, all Mitchell did was to indulge in a discussion of possible air tactics.[23]

Mason Patrick was embarrassed that Mitchell had gone outside his mission and had concerned himself with Summerall's command of the Hawaiian Department. "I am inclined to regard General Mitchell's report as a theoretical treatise on the employment of Air Power in the Pacific, which, in all probability, will undoubtedly be of extreme value some ten to fifteen years hence," he explained to Summerall. Mitchell, Patrick wrote, "has formed certain radical conclusions relative to the tactical employment of Air Force, in which, with the Air Force operating as a Branch of the Army, neither you nor I can fully concur." Perhaps as technology improved, Mitchell's views would be on target, but in 1923 what Mitchell wrote was speculative prognostication.[24]

Mitchell next went to the Philippines, where air defenses were truly deplorable. On January 12, 1924, he inspected Clark Field and found that replacement parts were almost nonexistent there and that the landing lights at the field were not hooked up because there was no electric current at night. Communications with Manila went through a commercial cable, and tires for the aircraft were affected by tropical rot. The machine shop truck could not be started, and there were no spare parts to fix it. The problems in recruiting competent soldiers for the Air Service surfaced in spades again. Mitchell questioned a Private Diddy, who was supposed to be a mechanic, asking him if he knew anything at all about an airplane engine. Diddy replied that he did not, because he had spent his time in an engineering shop. Not all was hopeless, however, in that a number of officers and non-commissioned officers knew their business, and when Mitchell looked at take-off times in a practice alert, he was pleased to see aircraft in the air within fifteen minutes (if an enemy would be so obliging). When Mitchell had a chance to address the Clark Field personnel he stressed tactics, especially concentration against an enemy and air gunnery simulating actual combat.[25] It was good advice, if the pilots could get their planes off the ground in time, and if those planes were mechanically sound enough and modern enough to actually fly against an enemy.

Three days later Mitchell observed a war game and tactical exercise, an alert against the "Orange" force, an enemy attacking from the north. (There was only one potential enemy to the north—Japan.) Also in attendance was Brigadier General Malin Craig, who was now stationed in the Philippines as the commanding officer of the Coastal Artillery District of Manila. The aircraft designated for the exercise were seaplanes, and they were aloft within an acceptable half an hour. The condition of the repair parts, tires, and spare engines was obviously poor due to the tropical climate and the periodic typhoons that damaged everything from runways to the one communications cable that connected the airfield to Manila. When Mitchell inspected the barracks he found that they were infested with bedbugs, but because he had once served in the Philippines he knew the nature of the place and problems involved with serving there.[26]

These were good, profitable inspections by Mitchell. Almost every military inspection by higher headquarters begins with the inspector assuring those about to be examined that he is there to help. In this case, Mitchell's prior service in the Philippines and knowledge of conditions there allowed him to speak with authority when he returned to Washington. Unlike the Hawaii visit, Mitchell's observations did not irritate General Craig, and there was no surprise report left on his desk when Mitchell departed. The Mitchells went on to Singapore and then India, where Mitchell got in his tiger hunt and Betty Mitchell killed a tigress herself. "We had killed so many animals during our last three days," Mitchell wrote, "that their pelts were not sufficiently dry to pack, so we had to spread them on top of the automobile truck that was carrying our baggage south."[27] While slaughtering a large number of India's animals, including magnificent tigers, Mitchell learned little or nothing about air defenses within Singapore or, for that matter, India, but then, Patrick did not really expect him to.

What Patrick did expect, however, was some expert observation on Japanese aeronautics and Japanese operations in Korea and in the home islands. That was fine for Patrick, but certainly not for the Japanese Imperial government, which rejected Mitchell's visit. The War Department had done a poor job of coordinating the trip in the first place, and Congress's actions dealing with the exclusion of Japanese from immigration had inflamed the situation. Mitchell, of course, was not responsible for either problem, but one can understand why the Japanese would not want an American general with the reputation of Billy Mitchell scrutinizing their air service. The U.S. embassy in Tokyo informed the State Department that Mitchell's visit had to be stopped. It would be looked upon, the diplomats warned, "with grave suspicion by the Japanese."[28] Davis intercepted Mitchell in China and told him that under no circumstances was he to try to enter either Korea or Japan until it was agreed upon by the imperial government.[29] Finally, on May 26, 1924, eight months after leaving the United States, Mitchell was told to book passage directly to Nagasaki, Japan, as a tourist and to await transport home to the United States. He was not to go to Tokyo under any circumstances.[30] While Mitchell did as he was directed, he observed what he could as often as was possible.

Mitchell and Elizabeth sailed for home, and by June they were settled in a new house in Washington. Patrick had handled the rift with the highly irritated Summerall quite well, with Summerall telling the chief of the Air Service, "I agree with all you say as to General Mitchell's attainments and dominant force as the developer of aggressive action in the Air Service. He has done much, and will do much more to make it the powerful arm which it is destined to become."[31] Little did Summerall and Patrick realize that in a little over a year both men would be deeply involved with Mitchell and another confrontation over air. Unknown to Mitchell, Mason Patrick was working on a plan of his own pertaining to a reorganization of the air forces for the defense of the United States. Mitchell, meanwhile, had his own writing projects that would propel him back into the national spotlight as a fighter in the ongoing air debate.

Prior to his honeymoon Mitchell had finished the manuscript for a book entitled *Winged Defense,* which turned out to be a rehash of Mitchell's ideas since 1920. As he himself admitted, "This little book has been thrown together hastily. . . . Its value lies in the ideas and theories that are advanced which it is necessary for our people to consider very seriously in the development of our whole national system."[32] The main thrust of *Winged Defense* was an argument for a unified Department of National Defense with separate departments of air, army, and navy, which meant that the air arms of both the army and the navy would be taken away from them and placed under one head. Predictably, Mitchell renewed his fight against the navy by citing the *Ostfriesland* and *Alabama* tests to show that "aircraft dominate sea craft." That really came as no surprise, nor did it cause much public comment from Admiral Moffett of the navy's Aeronautical Bureau when the book was released in the summer of 1925.

Mitchell wrote for two reasons: he was always in need of money and knew that a number of magazines would pay to publish what he wrote, and he was a polemicist who wanted to have his words in print, reaching the largest possible audience. In 1924 Mitchell came to an agreement with the widely read family magazine *Saturday Evening Post* to publish a series of articles on airpower and air defense. Though his ideas were not new to those who knew him, those concepts, which were well presented, would now reach literally millions of readers. His titles included such well-worn ones as "Aircraft Dominate Sea Craft" (January 24, 1925) and "How We Should Organize Our National Air Power" (March 14, 1925).

But there was a problem with Mitchell's writing: as a brigadier general and an assistant to the Chief of the Air Service, Mitchell was a member of the government and the army, not a private citizen. As far back as the Mitchell-Menoher controversy, Secretary of War John W. Weeks had warned Mitchell (and Menoher, for that matter) about publishing articles for profit. The initial difficulty arose with articles sold to the magazine *U.S. Air Service,* and since 1921 Weeks had not changed his opinion.[33] Weeks and Patrick wanted manuscripts for articles and books cleared first with them because of the serious nature of the unified air arm controversy and the bad blood between the army and the navy over all air matters. Neither Weeks nor Patrick would have cared very much had Mitchell written an article about hunting tigers in India, but *Winged Defense* and, certainly, the *Saturday Evening Post* articles were in a very different category.

An even more serious problem with the articles for the *Post* was that Mitchell, who had to be very well aware of Week's position, violated a cherished and very necessary military institution, the chain of command. The chain is the established line of communication between a subordinate and his superiors. For example, if a lieutenant commanding an infantry platoon has a problem, he does not go directly to his division commander. He first brings the situation to his company commander, and if the company commander is unable to render a decision they go to the battalion commander. If the battalion commander cannot deal with the problem, the lieutenant is sent to the brigade commander and then, if necessary, to try to see his division commander. On the other side of the coin, in combat, unless a very extraordinary situation exists, the division commander, normally a major general, does not issue orders directly to that lieutenant. Orders go from division to brigade to battalion to company and then to that infantry platoon commander. Without a clear chain of command there can be chaos, and this is true in peace as well as in war. According to Mitchell, the editor of the *Post* had spoken to President Calvin Coolidge about Mitchell writing the articles, and Coolidge had agreed that it was fine with him as long as Patrick agreed. On November 12, 1924, Mitchell received a letter from Coolidge to that effect, and Mitchell claimed that he went to see Patrick about it. By then, if Mitchell did indeed broach the subject with Mason

Patrick, the first of the articles was in press, and Patrick was over a barrel.[34] In early 1925 Mitchell claimed, "I therefore complied with both the President's verbal and written instructions."[35] He went on to argue that if the War Department wanted his *Post* articles stopped, it could have done so. Mitchell seemed to say that if the secretary of war wanted to interfere with freedom of speech and the press, he could have.

This was particularly difficult for Mason Patrick, who was in the process of preparing a major statement on the reorganization of the Air Service. He would go through military channels with his proposals, but a Billy Mitchell dominating headlines could muddy the waters considerably. Patrick was careful with words; there was an economy to his writing, and his memoirs, published in 1928, can seem disappointing because he does not elaborate. When Patrick submitted a copy of the reorganization plan for the air forces, it was well thought out and simply stated. But the simplicity could not mask his feelings when he submitted his document, through proper channels, to the adjutant general of the army:

> I am convinced that the ultimate solution of the air defense problem of this country is a united air force, that is the placing of all of the component air units, and possibly all aeronautical development under one responsible and directing head. Until the time when such a radical reorganization can be effected certain preliminary steps may well be taken, all with the ultimate end in view. . . .
>
> As an immediate and practical step toward efficiency and economy, the Army Air Service should be definitely charged with all air operations conducted from shore stations. Ultimately all air forces in the aerial defense of our coasts should be placed under one air commander and powers delegated to this commander that would permit the greatest latitude in the performance of his mission.[36]

When Patrick handed his plan to General Robert "Corky" Davis on December 19, 1924, he effectively marginalized Mitchell, though that was not his main or secondary purpose. Mitchell had been the loudest voice for a unified air service, for air defense, but now a full-blown, two-star major general, a man personally selected twice by Black Jack Pershing himself, was the chief and authoritative spokesman for the air. It was Patrick, not Mitchell, who had grown in the job, and it was Mason Patrick who—while Mitchell was off on inspections, tours, air races, and horse shows—had publicly carried the message of the Air Service.

Patrick looked with great interest at the finding of a board headed by Major General William Lassiter that was charged to look into the state of aviation. Its findings were predictable: the condition of American aviation was deplorable. The board also consisted of two experienced airmen:

Lieutenant Colonel Frank Lahm, who had commanded the 2nd U.S. Army Air Service in France in the closing days of the war; and Major Herbert A. Dargue, who had undertaken a number of studies for Patrick, including one on the status of lighter-than-air ships in the army.[37] Mitchell had tried to get dirigibles under the control of the army for coastal reconnaissance and defense, but the navy had resisted, considering Mitchell's actions to be a grandstanding power grab. Patrick, who was himself interested in the airships, wanted to approach the question in a reasoned manner. The development of thirty-one airships for army purposes was part of the Lassiter report.

The Lassiter Board's recommendations, including a ten-year plan for the development of army air, was accepted in principle by Secretary Weeks but was rejected by Secretary of the Navy Curtis Wilbur in early 1924. Because the board called for the army to get the lion's share of appropriated monies (40 percent of the total), Aeronautical Bureau head William A. Moffett was forcefully and publicly opposed.[38] Moffett also believed that this was the first step toward Mitchell's unified air force. Much work would have to be done to reconcile views, if possible, and at that point it was good to have Mitchell touring, inspecting, hunting, and honeymooning in the Orient so that Patrick could present an army position without press headlines and army-navy antagonisms exacerbated. Patrick would not sit idly by and await events, however. When he issued his annual report on the state of the Army Air Service, Patrick pulled no punches in deploring the state of aviation. The underlying theme of his report was that American aviation could not meet the demands of a crisis situation and was basically in the shape it had been in when the nation went to war in 1917 and had to rely on the allies for aircraft. Aircraft production was, in Patrick's words, "haphazard and unsystematic," and the Air Service simply lacked up-to-date aircraft.[39] So serious were Patrick's warnings of impending disaster that the *New York Times* carried an editorial citing Patrick's expert observation and calling for the government either to get into the business of making aircraft for the military or to appropriate enough money for the aircraft industry to expand and supply new and better planes.[40] Patrick then went to Capitol Hill and told a House committee that in reality, American aircraft were so antiquated that they could not sink a modern battleship because their bomb-carrying capacity was too small. The navy had just completed bomb tests on the hull of the incomplete battleship *Washington* and, though the Navy Department had refused to release the results, Patrick did say that new hulls could resist the small bombs that old aircraft carry.[41]

It was uncharacteristic of Mitchell to remain in the background as many of his long-held ideas were being discussed by others, but that's exactly what he did, turning his attention to getting settled with "Betty" and attending to personal matters. After being away from his stables for nearly half a year, he found his herd in need of thinning, and he decided to sell a number of horses.[42] In late July Mitchell began to experience a very sore throat, and army doctors ordered him to have his infected tonsils out, which he did.[43]

Mitchell and Betty had decided to find a home in northern Virginia, close to the capital, where they could raise horses and dogs and be close to others who loved to raise the fine show animals. They were looking at several places in Loudon and Faquier Counties, considered to be the heart of the Virginia "Horsey Set." Mitchell himself could not afford a Virginia country home among the beautiful rolling hills, but Elizabeth could, and she would support the purchase and refurbishing of a home. Mitchell continued to live as a well-to-do gentleman, spending almost $700 on a fine British-made shotgun, and acquiring many of his hunting and riding clothes from a Saville Row tailor in London.[44]

There was a thorn among the roses for Mitchell, however. His initial appointment as assistant chief of the Air Service had been for four years, and nothing guaranteed that he would be given the assignment again. He had angered Weeks and Patrick, and there was a new chief of staff of the Army who would have some input into Mitchell's situation: John Leonard Hines, an old friend of Black Jack Pershing, who handpicked Hines as the new chief. On September 14, 1924, with a smiling Pershing looking on, his protégé was sworn in. Hines had commanded the 4th Division during the war and ended at the Armistice as a corps commander.

Hines's assistant chief of staff, a very important decisionmaking position, was Brigadier General Hugh A. Drum, appointed in early November 1923.[45] Hines also appointed, at Pershing's strong suggestion, Major General Dennis Nolan as deputy chief of staff. Neither man had any love for Billy Mitchell, and they both lobbied for a new assistant to Patrick when Mitchell's time ran out. But Mitchell still had powerful political friends, and he had a large number of supporters in the press. To remove Mitchell would take a major incident.

A week before Christmas the potential for a major confrontation over air policy increased dramatically when Congressman Charles F. Curry announced that he had prepared a bill to create a separate department for aeronautics. All of the public discussion during the fall of 1924 motivated the congressman to draw up a bill,[46] which, though flawed, promised hot debate. Congress announced hearings to assert the true state of affairs and what would be needed to correct what Patrick and others had been pointing to for almost a year. This is exactly the venue Mitchell wanted—a congressional hearing with the press in full attendance. Secretary Weeks tried to forestall a public airing of air problems until there could be some agreement between the army and navy on what was possible. Weeks designated Mitchell's old foe Hugh Drum to present the army case at the hearings, which were to begin when Congress returned from the Christmas–New Year's vacation.[47] Mitchell, however, was determined to fight, to be heard, and to be covered by the press.

NOTES

1. Patrick to Pershing, 11 November 1919, in the John J. Pershing Papers, Library of Congress, Washington, DC.
2. Mason Patrick, *The United States in the Air* (Garden City, NJ: Doubleday, Doran and Co., 1928), 89–90.
3. Patrick to Carl Spatz, 24 October 1922, in the Carl Spaatz Papers, Library of Congress, Washington, DC, Carton 2. (Hereafter cited as the Spaatz Papers.)
4. Ibid., 17 November 1922.
5. Bane to Mitchell, 8 August 1921, in the George Hardie Collection, Golda Meir Memorial Library Archives, University of Wisconsin, Milwaukee, Carton 6, Folder 12.
6. Mitchell to Spatz, 1 June 1923, Spaatz Papers, Carton 3.
7. Spatz to Mitchell, 4 June 1923, Spaatz Papers, Carton 3.
8. *Detroit Free Press,* 15 August 1923, Hardie Collection, Scrapbook 6.
9. Roosevelt to Fullam, 1 February 1923, in the William F. Fullam Papers, Library of Congress, Washington, DC, Carton 6. (Hereafter cited as the Fullam Papers.)
10. *New York Times,* 27 June 1923, 8.
11. Ibid., 18 April 1923, Sec. VIII, 13.
12. Ibid., 27 June 1923, 21.
13. Ibid., 25 July 1923, 10.
14. Ibid., 27 July 1923, 12.
15. Sims to Fullam, 10 August 1923, Fullam Papers, Carton 6.
16. Adjutant General's Office, Orders Dated 29 September 1923, in the William Mitchell Papers, Library of Congress, Washington, DC, Carton 10. (Hereafter cited as the Mitchell Papers.)
17. Clippings Files, Hardie Collection, Carton 10, Folder 8.
18. Lovell to Mitchell, 4 December 1923, Mitchell Papers, Carton 10.
19. Ibid.
20. National Archives, Archives II, College Park, MD, Headquarters, Hawaiian Military Department, Activity Report, November 2, 1923, in Records Group 153, Records of the Judge Advocate General, General Courts-Martial, William Mitchell Case Records, 1925, Case No. 168771, Entry 40, Carton 9214-12. (Hereafter cited as the William Mitchell Case Records, 1925.)
21. Ibid., January 3, 1924.
22. Summerall to Patrick, 27 December 1923, William Mitchell Case Records, 1925, Entry 40, Carton 9214-12.
23. Ibid.

24. Patrick to Summerall, 26 January 1924, William Mitchell Case Records, 1925, Entry 40, Carton 9214-12.

25. Clark Field Inspection Notes, 12 January 1924, William Mitchell Case Records, 1925, Carton 46.

26. Kindley Field Inspection Notes, 15 January 1924, William Mitchell Case Records, 1925, Carton 46.

27. William Mitchell, "Tiger-Hunting in India," *National Geographic* 46, 5 (November 1924), 598.

28. Tokyo Embassy to State Department, 16 May 1924, Mitchell Papers, Carton 10.

29. Davis to Mitchell, 17 May 1924, Mitchell Papers, Carton 10.

30. U.S. Consul General at Mukden to Mitchell, 26 May 1924, Mitchell Papers, Carton 10.

31. Summerall to Patrick, 18 February 1924, Mitchell Papers, Carton 10.

32. William Mitchell, *Winged Defense* (New York: G. P. Putnam's Sons, 1925), viii.

33. *New York Times,* 12 June 1921, 3.

34. Mitchell to Morrow, 2 March 1925, Mitchell Papers, Carton 46.

35. Ibid.

36. Memorandum by Patrick for Davis, 19 December 1924, Mitchell Papers, Carton 10.

37. Memorandum from Dargue to Patrick, 4 November 1924, Mitchell Papers, Carton 45.

38. William F. Trimble, *Admiral William A. Moffett: Architect of Naval Aviation* (Washington: Smithsonian Institution Press, 1994), 141–142.

39. *New York Times,* 21 November 1924, 6.

40. Ibid., 18.

41. Ibid., 14 December 1924, 24.

42. Mitchell to Michael Cudahy, 31 July 1924, Mitchell Papers, Carton 10.

43. Ibid., 8 August 1924.

44. Colonel Kenyon A. Joyce (Military Attaché, London) to Mitchell, 21 October 1924, Mitchell Papers, Carton 10.

45. *New York Times,* 7 November 1923, 17.

46. Prepared Statement by Curry, ca. 20 December 1924, Mitchell Papers, Carton 31.

47. Weeks to Congressman John C. McKenzie (Chairman, House Committee on Military Affairs), 8 January 1925, Mitchell Papers, Carton 37.

NINE

GLORY, GLORY, BILLY MITCHELL

N O ONE IN THE WAR DEPARTMENT, THE OFFICE OF THE chief of staff of the army, or the office of the chief of the Air Service believed that Mitchell would be reticent, a team player, in the upcoming hearings, and some were counting on his speaking out. The Honorable Mr. Charles Curry (R-Calif.), had been introducing bills for a separate air service since 1919, and hearings on another Curry bill were scheduled for late January 1925.[1] One of the early witnesses was Brigadier General William Mitchell, Mason Patrick's assistant, and the appearance, without his consulting Patrick as to testimony, promised to have real fireworks. The House committee was headed by Congressman Julian Lempert of Michigan, and included a number of Mitchell's longtime social and political friends, such as Frank R. Reid, who represented a congressional district from Illinois. Most of the Democrats on the committee knew that Mitchell's grandfather and father had been Democratic Party officials and that Mitchell was a self-proclaimed member of the party of Jefferson and Jackson. There was even some talk that Mitchell might resign from the army to run for the Senate from his home state of Wisconsin.

As soon as Mitchell took the witness chair on January 18, 1925, he set the tone for his testimony with a blistering attack on the navy's experiments on the hull of the *Washington*. Both Mitchell and Patrick had watched the experiments, and the two men came away with different views. Mitchell told the committee that the navy really had used no bombs, only underwater explosions, and their findings were phrased so as to purposely deceive the nation. Patrick had publicly stated that the bombs used against the *Washington* were too small given the type of antiquated aircraft the air services had to contend with.[2] Old aircraft simply could not carry larger, more effective bombs. Somebody was not telling the truth. Mitchell's hostile attack caused a number of his most ardent supporters to have second thoughts. Major Henry "Hap" Arnold, who was serving in Mason Patrick's office as information director, confronted Mitchell and urged him to be careful of attacking senior army and navy officers in public. Mitchell, now in a state of high excitement, shrugged off Arnold's warnings, saying what he was doing was for the good of the air services and the nation.[3] Billy Mitchell, a professional soldier since 1898, was transgressing military protocol in a most public way, and the army, the navy, and the Secretary of War were furious; their faces got even redder as Mitchell made headlines in every major newspaper in the country.

Newspaper reporters and subsequent biographers have tended to focus on the Mitchell testimony, which was explosive. But some nasty political games also were being played at the hearing, and Mitchell was a part of them, knowingly or unknowingly. Frank Reid of Illinois, who later was one of Mitchell's counsels at his court-martial, had been making his own political hay during the hearings. Reid was a Republican maverick from Aurora, Illinois, who had been a successful attorney prior to his election to the House of Representatives in 1911. In February 1923 Mason Patrick published an article in *Current History Magazine* entitled "Cost of Our Wartime Aircraft," and he cited the figures for what the United States had spent to support the war effort in the air, in combat and training.[4] Reid announced that the costs were much higher for the period between 1920 and 1924, and he went on to claim, "This is the real 'aircraft scandal' of which we have heard so much. In the past charges have been made of graft and industrial deficiency and concerned mainly the strictly war time period ended June 30, 1919. Now we find that in the postwar period from July 1 to date, our aviation expenditures have actually been greater than during the war."[5] Reid claimed that he was not charging graft, but he intended to find out where that money went, implying that higher-ups in the War Department, the General Staff, and the Air Service might indeed be responsible. Reid's questions for Mitchell were designed to allow Mitchell the greatest possible stage to make his own claims.

Mitchell had a tendency to speak in generalities, with wide, sweeping claims or charges, and he was in fine form during the hearings. At one point Mitchell stated that air policies were tied to established governmental agencies that really had no concern about the air. "The result is that all the organization that we have in this country really now is for the protection of vested interests against aviation," he told one congressman. A member of the board was perplexed and wanted Mitchell to explain his statement with specifics. Mitchell responded with some talk about the army, navy, and post office, and then stated, "The personnel for the Air Service is selected from people who have been trained for other objects first and then for the air second. It is a waste of time to train a man to dig a hole in the ground to get away from hostile shellfire when he is in the air force and must fight 20,000 feet up in the air. That is about what we have been doing." Growing more exasperated with Mitchell, the congressman asked if the United States was the only modern nation in the world to have such an organization, to which Mitchell responded, "Yes, practically the only one." What, then, was the reason for this state of affairs, Mitchell was asked, and he responded, "Conservatism . . . in the Army and Navy."[6]

Billy Mitchell, in full uniform, with his walking stick and his rows of colorful ribbons, was not through. When asked about air needs, he responded, "It is not a question of money either. I think the total amount of money being put into aviation is plenty." When Congressman Perkins pushed Mitchell on his astounding statement, Mitchell indicated that the sums granted by Congress were enough if properly administered. Billy Mitchell then cited an unnamed friend who had asked him what it would take for the United States to gain control of the air, and said that he told this fellow, "if we could get half the cost of a battleship each year we could control it in two years." The trouble, then, the congressman summarized, was not with the congressional appropriations but with the spending of the funds by the military. Mitchell responded that he did not think the problem rested in the halls of the Congress.[7]

Secretary Weeks was in a full rage over Mitchell's strong hits against the armed services. Mason Patrick, although he did not say so at the time, had to have been personally hurt when he recalled the numerous times he had presented strong cases to the House and the Senate for increased funding for the Air Service. Mitchell's testimony was a direct slap at Mason Patrick, who had protected him for three years. Billy Mitchell had also played into the hands of Generals Nolan and Drum, and Admiral William A. Moffett of the navy's Aeronautical Bureau. The newspapers were having a field day with Mitchell before Congress, and the *Washington Herald* of January 19 carried the headline "Aviation Chief Scores Army and Navy Autocrats." Weeks had to act, and he prepared the adjutant general's office to comb the Mitchell testimony for objectionable parts, put them into a memorandum for General Patrick, and have Patrick order Mitchell to respond to each section.[8] In reality this was a formality, because Weeks had decided that Mitchell could not remain assistant chief to Mason Patrick. Mitchell's appointment as the assistant was due to expire in a few months, and now Weeks had the ammunition he needed to sink Mitchell.

Mitchell obliged Weeks with even more rounds to fire back at him, because much of the controversy was driven now by politics and political figures. Congressman Randolph Perkins of New Jersey knew exactly how to elicit the best responses from Mitchell. The flamboyant and highly intelligent Fiorella LaGuardia of New York had continually supported the advancement of American air since the end of the Great War, and he saw in the hearings a chance to take advantage of the press and public attention to push his cause. Like any successful politician, LaGuardia grabbed the question of the *Washington* experiments and made it his cause célèbre during the hearings. LaGuardia was delighted with Mitchell's attacks against the navy and used them as a starting point for his own assault. He accused the navy of refusing to give the hearings any specific information. He lashed out at Navy Secretary Curtis D. Wilbur, stating that congressmen were "now confronted with the startling situation that we can not obtain accurate information as to the results of the bombing."[9] He said that Mitchell must be recalled to the halls of Congress to give further testimony, and Randolph Perkins and Frank Reid quickly agreed. Mitchell was now being used, and he was happy to be in that situation with a friendly audience and an eager press.

On February 4 General Drum appeared before the committee to state the army's position on the Air Service. As with Mitchell's testimony, there were no new positions stated, no new ideas presented, and it was again a political event. Drum simply stated that the air was a vital support for ground operations, where battles were won and lost. There was no reason for a separate air service because all army branches had to work together for final victory. Then Drum recalled the Great War, especially the Meuse-Argonne campaign where Mitchell had placed great emphasis on hitting enemy air fields to keep the German Air Service from flying. "Nothing of the sort happened," claimed Hugh Drum, because reports had reached army headquarters of German planes flying over American troops at will. In closing, Drum stated that no high-ranking aviation officer had ever commanded a combined arms team in battle. They knew only one aspect of war fighting, and that was not the way to win wars.[10]

Drum was fully prepared to defend the official policies of the War Department, but he was not ready for political probing by the congressmen. Randolph Perkins asked Drum, "Do you think that the possible removal of Admiral Moffett and of General Mitchell from their present position, as it is rumored will happen, will have the effect of preventing army and navy officers from talking frankly to committees of Congress?" Drum was confused, and hesitated in his response, finally saying, after an embarrassing silence, that he did not think that it would, and, at any rate, Secretary of

War Weeks had a fairly lenient policy when an officer expressed personal views.[11] Admiral Moffett did have his concerns about being reappointed as head of the navy's Bureau of Aeronautics and was ready to fight for his position, especially since Mitchell seemed determined to throw his away.[12] Moffett denied Mitchell's contention that officers were coerced by the Navy Department and that the War Department was simply wrong.[13] Navy Secretary Wilbur also sent a letter to the Lempert Committee denying that any naval officer had ever been silenced by superiors.

When Mitchell returned to the stand he was ready to go further than he had a few days before. He immediately went after Hugh Drum: when Perkins asked him about what Drum knew about the Air Service, Mitchell replied, "Nothing whatever." Mitchell charged that members of the War Department and Navy Department had deliberately distorted facts when they testified before Congress. "If any civil officer should be found guilty of such distortion," Mitchell then stated, "he should be impeached and if a naval or military officer he should be court-martialed." Billy Mitchell told the committee that upon his return from his inspection tour of the Pacific he had filed a report that no one had dealt with. Reporters quickly learned from Mitchell's supporters that the report pointed out the deplorable condition of the defenses in the Pacific and the international dangers in that area, and that it would create a sensation when released.[14] The report had been sent to the War Department and then filed under the control of the adjutant general—there was then no sense of national urgency about what were basically routine inspection reports. The adjutant general of the army, Major General Robert "Corky" Davis, saw no need in 1924 to push for a full-scale review and response by the General Staff. For congressmen, journalists, and the public who followed the testimony on Capitol Hill these reports took on a significance that was out of proportion to their actual value at that time.[15]

This was an unfair, low blow by Billy Mitchell because his report had not been completed and filed until the fall of 1924, and it was well over three hundred pages long. Mitchell knew full well that the General Staff needed time to study such an in-depth report and to formulate responses. Normally, when such reports are received by a staff they are broken down so that each section, especially the G-2 (intelligence), G-3 (operations and training), G-4 (logistics), and the Air Service, could study and respond to what the report contained by either agreeing with sections and offering suggestions to correct problem areas or countering claims made. That meant that those sections had to collect their own information. The Mitchell Pacific report was filed in peacetime, when there was not as great a sense of urgency as there would have been in wartime. Weeks was caught unaware by Mitchell's claims, and stories by reporters made it appear that the staff was hiding the reports and that Weeks really was not being fully briefed. As it was, the staff would complete their detailed response to the report in 1926.

To complicate matters, Mitchell told the committee that on December 10, 1924, Mason Patrick had sent the adjutant general a memorandum that called for a reorganized, unified air service. The *New York Times* reported that this memo was revealed for the first time.[16] By this time Weeks was beside himself with anger, and he directed the adjutant general to have Mitchell respond to yet another inquiry, his second one. This time Mitchell was to give specifics, and Patrick was to recommend what should happen to Mitchell for these serious breaches of military conduct.[17] On February 11 the adjutant general's office, moving toward some sort of legal action against Mitchell, directed Patrick to have Mitchell respond to several specific accusations. One matter that aggravated the army was that Mitchell had testified he had told officers within the Air Service to remain silent "and let me assume all responsibility." It appeared to the lawyers that Mitchell had given an "official statement to a Congressional committee, which he knew was contrary to explicit and personal instructions of the Secretary of War." To tell officers to be silent and allow Mitchell to shoulder the consequences was in direct violation of War Department policies, and Mitchell "endeavored to create in the minds of a Congressional Committee, the public at large and officers of the services an erroneous impression of the War Department's policy relating to officers testifying before a Congressional Committee."[18] The tone of the directive to Patrick was clear—the army was considering legal action against Mitchell.

Mitchell responded in a defiant tone; and he replied that yes, officers had come to him and said that if they testified they feared disciplinary action by the War Department. If the adjutant general wanted some names he would provide them in confidence, but he would not name them publicly. As far as his acceptance of responsibility, that was true. Secretary Weeks had once told him that if he could not support policies, he should resign. He then denied that he had ever received any restrictions on testifying or publishing. "I deny," he went on, "having made an official statement contrary to the instructions of the Secretary of War. I merely stated facts." As far as Mitchell was concerned his actions were open and aboveboard, and the questions being posed to him were basically trumped up and baseless charges.[19]

Much now would depend on Mason Patrick, Mitchell's superior officer, who would add his own views as to his subordinate's conduct. Surprisingly, Patrick defended a number of Mitchell's actions though he must have known it was a lost cause. While Patrick could not find any written instructions as to testimony, he did state:

> I called him into my office and did tell him I was quite sure he was authorized when called as a witness before a Congressional Committee to set forth his views, his opinions, fully and freely. I added that in so doing he should be sure of his statement of fact. General Mitchell replied in such a way as to make it apparent to me that he understood correctly and clearly what I had told him.
>
> I believe that the way in which General Mitchell gave his testimony and particularly such publicity was given it did tend to create in the minds of the people the impression that the War Department was unwilling for officers to testify freely.
>
> I recommend that General Mitchell be admonished by the War Department for his attitude and methods in the premises.[20]

The handwriting was on the wall for Mitchell. Eddie Rickenbacker, always Mitchell's friend and supporter, telegraphed him on February 19, 1925, "A man of your courage, initiative and intelligence is needed in business which will pay you more than you can ever hope to receive from Government. . . . I herewith tender you executive position Rickenbacker Motor Company at salary higher than you are now receiving plus full appreciation [of] your efforts and ability."[21] The same day, Mitchell wired back, "Very many thanks for your kindness. . . . I have my work pretty well cut out for next two or three years until we get a separate Air Service."[22] Either this was sheer bravado or Mitchell had lost complete contact with reality and had come to a point where he believed what reporters were saying about his testimony. In 1921, when his difficulties with Caroline Stoddard Mitchell had reached a crisis, both Mitchell and his adoring sister Harriet believed that no one, including General John J. Pershing or General James Guthrie Harbord, had the courage to deal with him. His testimony before the Lempert Committee and his response to the memoranda sent by the adjutant general were similar to his attitude toward his superiors in 1921. It was a disturbing pattern that did not bode well for the future.

On March 4 Weeks acted, as he had planned to do for some time. In a letter of explanation to President Calvin Coolidge, Secretary Weeks stated that he would not appoint Mitchell to another term as assistant to the chief of the Air Service. He went on to say Mitchell's "whole course has been so lawless, so contrary to the building up of an efficient organization, so lacking in reasonable team work, so indicative of a personal desire for publici-

ty at the expense of everyone with whom he is associated that his actions
render him unfit for high administrative position such as he now occu-
pies."[23] President Coolidge, who was personally angered over the public
spectacle and disgusted with Billy Mitchell, agreed with Weeks. Now the
question arose as to where to send Mitchell, who would go as a colonel.
Unlike Hugh Drum and others who were permanent brigadier generals,
Mitchell's rank was only temporary, depending on the job as assistant.
Once Mitchell left the job he would revert to his permanent army rank.
Rumors began to fly about Mitchell. The *Milwaukee News* reported that
Congressman LaGuardia was spearheading a move to force Billy Mitchell
in as chief of the Air Service. The *Washington Star* reported that Mitchell
had requested assignment as head of the Air Service for the army's VI
Corps, headquartered in Chicago. That the army would not do, because it
would give Mitchell full access to the Midwestern press. The orders issued
to Mitchell assigned him to the VIII Corps at Fort Sam Houston, Texas,
which was a large and important army post with a great deal of air service
training.

While Mitchell was making headlines with his testimony to the
Lempert Committee, Elizabeth, who was now pregnant, was moving
household items into the northern Virginia home she had just purchased. In
late 1924 the Mitchells had located the home that they wanted to purchase
in the horse country of Virginia, situated in beautiful rolling hills with the
gray outlines of mountains in the distance. The home was known as
Boxwood and had been built by William Swart in 1826. Constructed from
stone and situated on a small rise, the estate got its name from the boxwoods in the gardens.[24] It sat on an estate of 120 acres of farmland that
was well suited to raising horses and dogs and enjoying the life of the
country gentleman and family.[25] Elizabeth immediately began a renovation of the building by gutting it and adding a two-story addition to the
side and an elaborate entrance door. Mitchell then applied for a two-month
leave to take Elizabeth to Detroit, where their daughter Lucy would be
born and where Elizabeth would remain while Boxwood's rebuilding
could be completed. Before going to Fort Sam Houston, Mitchell visited
Meadowmere and selected furniture to send to the home near Middleburg,
Virginia.

Mitchell was replaced by Lieutenant Colonel James E. Fechet, who
was immediately promoted to the job as a brigadier general. There had been
gatherings for Mitchell before he left Washington, but few senior officers in
the War Department were genuinely sorry to see Mitchell go. Circulating
around the staff was a lengthy parody to the tune of the "Battle Hymn of
the Republic" which went in part:

I'm Billy Mitchell, I'm the Eagle of the Lord
I have rendered weapons obsolete from Battleship to sword
Ideas rattle in my head as seeds do in a gourd
AND I FLAP MY WINGS AND CROW
Glory, Glory Billy Mitchell
Glory, Glory Billy Mitchell
Glory, Glory Billy Mitchell
I flap my wings and crow
I can make a thrilling statement, and am careless if it lacks
any elements of truthfulness about my senior's acts
I'm a man of strong OPINIONS, so I don't need any facts
AND I FLAP MY WINGS AND CROW
(Refrain)
I can sit way down in Texas and unerringly can note
any trouble had by Navy men with sea plane or with boat
So I write 'em up and thereby catch the bolshevistic vote
AND I FLAP MY WINGS AND CROW
(Refrain)
I would destroy the Army and I'd sink the Navy, too
to get a separate Service for the airplane and its crew
PROVIDED that the generalship to Billy would accrue
AND I FLAP MY WINGS AND CROW.[26]

No one believed that Mitchell could be quiet for long. When he left Washington for San Antonio, Texas, he gave the distinct impression that he believed he was right and the army and the War Department were wrong. In Texas he had no one like a Mason Patrick who could even make an attempt to calm Mitchell down and force him to devote his time to the Air Service within the VIII Corps. Without consulting any superior officer, during the summer Mitchell contacted John N. Wheeler, executive editor of the weekly magazine *Liberty,* and offered to write six articles on aviation. Each article would bring Mitchell a handsome $1,500, and to sweeten the deal Wheeler promised that Mitchell would receive the magazine's "hero award." The thrust of what Wheeler wanted was clear enough when he told Mitchell, "I have the highest regard for your attitude, the way you conduct yourself, and respect your views. . . . One article I would suggest is WHAT'S THE MATTER WITH OUR AIR SERVICE? In this you could express your views and tell how many of our aviators were lost in France because of inferior machines, how many were lost since, and so on."[27]

Mitchell's first article for the magazine was another blast at the navy and the battleship. In July Mitchell, on an inspection tour to San Diego, California, made a fiery speech about the deplorable condition of the United States Air Service.[28] No, Billy Mitchell would not accept his Texas assignment as a good soldier should, and, no, he would not be stilled. In July he published an article on arms limitations, again focusing his attention on the battleship, which he claimed was too costly for any nation to maintain because it was obsolete.[29]

On September 3, 1925, word reached Mitchell that the navy's dirigible, the *Shenandoah,* commanded by an old friend, Commander Zachary Landsdowne, had crashed in southeastern Ohio with heavy loss of life, including Landsdowne's. During the bombing attacks on the *Ostfriesland* Mitchell had made sure that Lansdowne had a spot on the USS *Henderson* to observe the action, and it is quite likely that Mitchell was personally distressed by Landsdowne's death. Zachary Landsdowne and his wife had seen the Mitchells socially, and Betty Mitchell had befriended the attractive Margret Ross Landsdowne. That relationship immediately set off alarm bells in the Navy Department. Given the bad blood between Mitchell and the navy which had existed since 1919, there was no reason to believe that he would remain silent. It would have taken the intervention of the Almighty for Billy Mitchell to make general officer on a permanent basis, so there was no reason for him not to speak out when the tragedy also included the death of a friend, and speak out Billy Mitchell did.

On September 5, Mitchell called six local reporters to his office at Fort Sam Houston and gave them a six-thousand word statement. Many of the copies were hard to read since they were carbon copies. Mitchell cited only part of the statement, saying, "These accidents are the result of the incompetency, the criminal negligence, and the almost treasonable negligence of our national defense by the Navy and War Departments." Much of what Mitchell wrote was vintage Billy, but the charges were bound to bring on a rapid and sever reaction from Washington. Mitchell railed about the *Shenandoah* being an experimental aircraft when, in fact, the year before the ship had made a transcontinental flight. Predictably, the navy was furious and the War Department was equally angered, ordering the corps commander to investigate and relieve Mitchell from duty. Admiral Moffett, that longtime foe of Mitchell's, was in San Francisco when he read of Mitchell's charges, and he was enraged. The old salt and director of the navy's air component told a fellow officer, "That son-of-a-bitch is riding over the Navy's dead to further his own interests."[30]

Mitchell's outburst came at a bad time for the War Department, because Mason Patrick had just been reappointed as chief of the Air Service in late July. His term was to expire in early October, but Weeks, who himself was ill and on the verge of retiring as Secretary of War, wanted stability after the Lempert hearings and with new hearings in Congress about ready to start.[31] Patrick was a stabilizing factor for the younger aviators, who looked on Mitchell as their firebrand but also saw Patrick as, "our great leader. . . . He had the great faculty of being able to talk to the military leaders, the Chief of Staff of the Army, the chiefs of the other services, the Secretary of War, the Assistant Secretary of War, and the senior members of the Congress."[32] Billy Mitchell's conduct was considered to be so outrageous that something would have to be done, and it added a great burden at a time when stability was needed. There were questions as to what exactly Mitchell did say and write and what were reporters' comments. The commander of the VIII Corps was instructed to find out what actually had been said. Lieutenant Colonel George K. Hicks, adjutant general of the corps, sent Mitchell a copy of the article printed in the *San Antonio Light* on September 9, 1925, and asked him whether that was indeed what he had said and wrote. Hicks ended his request by telling Mitchell that he could remain silent as a constitutional right because any statement he might make could be used against him.[33] Responding that he had indeed made those statements,[34] Mitchell understood that he was seeing the first phase of a court-martial. Mitchell was relieved of duty and was told to stand by for orders that would send him back to Washington.

The Mitchell controversy was complicated by politics, and Billy Mitchell was up to his neck in politics and enjoying it. After his dramatic testimony before the Lempert Committee and his posting for Fort Sam Houston, Texas, his reputation had become nationally known. Conditions in Wisconsin were unsettled by the death of long-time Progressive Senator Robert M. La Follette on June 18, 1925. The governor of Wisconsin had appointed La Follette's son as the new senator, but there were other opportunities in Mitchell's home state. Key members of the Democratic Party were interested in Mitchell as a candidate for some position in 1926. Political operatives offered to work for Mitchell for a large salary to line up support within the party for some sort of Billy Mitchell candidacy.[35] Billy Mitchell had always talked about a political career from Wisconsin after he retired from the army, and he certainly felt that if a Cabinet position for air was created he should be the one to head it in a Democratic administration. Consequently, as the army began to prepare court-martial charges against Mitchell they had to be aware of the political nature of a trial.

While Mason Patrick and John Leonard Hines were testifying before a special board called by President Coolidge, Mitchell arrived in Washington and was told in no uncertain terms that he was to be quiet. In mid-September the adjutant general of the army began preparing a charge sheet against Mitchell. This was a legal document spelling out exactly what Mitchell was to be tried for. Once completed, it would be given to the accused in time for him and his counsel to prepare a defense. The army would provide legal counsel for Mitchell, and, if he so wished, he could have a civilian lawyer defend him at his own expense. The sheet was com-

pleted and signed by the adjutant general on September 24, 1925, and was
sent to Mitchell and to Colonel Sherman Moreland, trial judge advocate of
the army, who was responsible for setting a date for the trial at a suitable
location. The charge was a simple one: "conduct to the prejudice of good
order and military discipline and in a way to bring discredit upon the military service." The six journalists from the San Antonio area whom Mitchell
had called to his office to hear his statement and receive copies of his
lengthy written statement were listed as potential witnesses against
Mitchell.[36] The final document was a massive fifty-five pages because each
individual reporter's copy of the six-thousand-word statement was included
as seven individual specifications under Article 96 of the army's Articles of
War, the army's standards of conduct.

Mitchell left Fort Sam Houston by train for Washington, and Betty
boarded the train in St. Louis. When they arrived in Washington on
September 25 the couple was greeted by a raucous crowd at Union Station.
The demonstration was orchestrated by the American Legion, which turned
out a crowd of several thousand yelling, placard-carrying supporters.
Mitchell, who was not under military arrest yet, checked into that fine,
plush Washington landmark, the Willard Hotel, until larger accommoda-
tions could be found to prepare for the trial. The next day Mitchell went to
a Legion-sponsored barbecue, where he was again surrounded by a throng
of admirers. This very public adulation and ballyhoo was not appreciated
by all, however. One Washington editorial writer told his readers, "Colonel
Mitchell's testimony will be somewhat delayed as it is understood that all
arrangements for the movie rights have not been completed."[37]

Billy Mitchell was far from languishing in his surroundings while he
pondered his defense. Not surprisingly, Mitchell called on Illinois
Congressman Frank Reid to be his lead civilian defense counsel when the
court-martial began. While he was pouring over papers, the Air Board
Inquiry went on capturing headlines, dominating the public's interest for
the month of September. The officers who went up to Capitol Hill read like
a Who's Who of the AEF in France, including some of the most experienced battlefield commanders in the United States Army. Assistant
Secretary of War Dwight Filley Davis orchestrated the impressive military
display, conferring constantly with Chief of Staff John Leonard Hines and
Hines's assistant Hugh A. Drum, who was selected to give the army's presentation at the end of the inquiry.

Hines, Drum, and Davis knew what the press did not—John W.
Weeks's health was rapidly deteriorating and Coolidge had decided that
Davis would replace Weeks before Thanksgiving 1925. For Mitchell the
change would have special importance, because Davis was even more

determined to defend army interests than was Weeks. Dwight F. Davis, born in Missouri, had graduated from Harvard and attended Washington University of St. Louis' School of Law. An avid tennis player, Davis established the Davis Cup for excellence in tennis play, but war pushed his athletic interests to the background. He was a graduate of the Plattsburg, New York, officers program, had served in the AEF in the 138th Infantry of the 35th Infantry Division, and saw heavy combat during the Meuse-Argonne offensive. He served with an obscure artillery battery commander by the name of Captain Harry S. Truman. He was not a man who would allow Billy Mitchell to attack the army or to thumb his nose at army regulations and directives.

To try to sort out the confusion over air matters, President Coolidge appointed a special board headed by businessman Dwight Morrow. On September 21, Mason Patrick testified before the Morrow Board, defending the Air Service. When asked if Billy Mitchell's claims that Air Service officers were transferred throughout the army with no reference to Patrick, he stoutly denied it. When asked if he supported the idea of a separate air force, Patrick restated the view he had given the adjutant general in December 1924, which was an unqualified yes.[38] In the interest of doing things correctly, Patrick did say that while he favored the unified force and a Cabinet-level position overseeing the development of air, he believed that it was wise to go slowly. To rush into a new configuration affecting the defenses of the nation would be to invite chaos with no real progress. John Hines followed Patrick and was questioned by Congressman Carl Vinson of Georgia, who queried Hines about requests for legislation and money for the Air Service. Hines did not handle the questions very well, and when asked by Vinson if the army planned to submit any requests for legislation and funding, Hines could only say that he hoped so, perhaps by December.[39] The real star of the army's presentation was Hugh Drum, who very quickly took on Mitchell's assertions about the Air Service being a stepchild of the infantry. Quite the contrary was true, Drum told Dwight Morrow, chairman of the board of inquiry; the army had to train and work as a combined arms team. The idea of a separate air arm was foolish when one considered the need for all branches to work in tandem to defeat an enemy.[40] The Great War, Drum proclaimed, had proven this to be accurate, and to deviate from combined census was to invite defeat.

When Hines left the Morrow hearings he was beset by reporters who wanted to know if it was true that Mitchell had been relieved from duty and ordered to Washington. The army chief of staff said that it was true, and that he had issued the order to preserve discipline, not to punish Mitchell. In his view Mitchell's statements bordered on insubordination, and if

Mitchell were allowed to remain in command of VIII Corp's Air Service, which demanded obedience from his subordinates, it would send the wrong message about how the army viewed good military order. Being very careful in his response, Hines said several times that this order of relief was not a disciplinary measure, and, reading between the lines, the reporters could gather that indeed the army was planning to court-martial Billy Mitchell. Mitchell, however, stated that he expected to be called before the Morrow Board so he could defend the charges of negligence and near-treasonable activities.[41]

To be fair in this highly charged political atmosphere, the Morrow Board had to call Mitchell to testify, and they wanted to do it quickly because every member of the board expected fireworks.[42] Sitting on the board was retired Major General James Guthrie Harbord, who as Pershing's deputy in 1921 gave the orders sending Mitchell to the psychopathic ward at Walter Reed Army Hospital. Harbord had retired from the army to take a very important and lucrative post with the RCA Company. Having a keen memory, Harbord also recalled the difficult times the AEF had with Mitchell in France during the war. Another member of the board was retired Rear Admiral Frank Fletcher, who had openly criticized Mitchell for his grandstanding during the *Ostfriesland* bombing trials in 1921.[43] Both men warned Dwight Morrow that Mitchell could be expected to be difficult and theatrical when he appeared on September 29. Mitchell was to follow Benjamin Foulois, an old antagonist who, ironically, held the same views in regard to a unified air force as did Mitchell, but unlike Mitchell was guarded in his public statements.

Billy Mitchell's appearance began with Morrow stating that Mitchell would not be sworn in as a witness and could give his full testimony without interruption for questions. What Mitchell did was to read chapters from *Winged Defense* for hours in a dull monotone. Mitchell's wife and supporters were distressed that an opportunity to express his views in the old Mitchell way was lost, and Major Henry "Hap" Arnold wished silently that Mitchell would just close up the book and talk directly to the committee. Mitchell had not been formally charged by the army at that point, nor had he been placed under arrest pending a court-martial, and he was basically a free agent to say what he wished as Patrick, Drum, Hinds, and Foulois had done. Mitchell returned the next day, but there were no explosions in the room. While Mitchell's friends were distressed at his performance, they were seeing reality. Billy Mitchell had nothing new to say, nothing startling to add to what was transpiring at the Morrow Board. He had said everything he had to say and had reached a point where only the theatrics were left, and they did not materialize.

That was not how Billy Mitchell's supporters saw him, however, and to them he was still the great fighter against entrenched interests in the Navy and War Departments. Regardless of his lackluster testimony before the Morrow Board, Mitchell knew how to mobilize those who were his fans. From the Willard Hotel Mitchell wrote to Eddie Rickenbacker, sending him a copy of part of the statement he made to the board, which he wanted circulated among the delegates to the American Legion Convention in Omaha, Nebraska, in early October. He urged the great ace of the 94th Aero Squadron to go to the convention and read his demands for a unified air service: "What we [meaning the Legion and Mitchell] want is a Department of National Defense, under a Secretary, with land, air and water under it; the Navy to keep anything they want but they must go to sea with it; the army to keep anything they want for observation purposes; the Air Force to handle all military aviation and the Department of the Air to handle civil and commercial aviation and other aeronautical development. This will operate to fix the responsibility on one man, the Secretary of National Defense."[44]

The members of the Morrow Board, the War Department, and the Navy Department watched with dismay as the American Legion heard Mitchell's statement and reacted with a tremendous ovation of support. Admiral Moffett and other naval aviators were concerned that the whole process, starting with the Morrow Board and the Navy's court of inquiry into the *Shenandoah* crash, could be upset by this political activity by Mitchell and his friends.[45] Only one conclusion could be drawn from all of this—if there was to be a Secretary of the Air or a Secretary of National Defense, Billy Mitchell believed that he should be it. It was time the army took steps to take some of the wind out of Mitchell's sails, if it was not too late.

When Mitchell arrived in Washington he requested three officers from the Air Service to compile data for the Morrow Board. Captain Robert Oldys and First Lieutenants Clayton Bissell and David G. Lingle were assigned to Mitchell on temporary duty, and on October 2 Mitchell requested that they remain with him for two or three more days. When October 7 dawned and the three officers were not back at their desks, Brigadier General Samuel D. Rockenbach, a tank officer from the AEF and now commanding general of the District of Washington, asked the adjutant general to send an order to Mitchell telling him to return them to their duty assignments that very day. "It appears that the necessary time to prepare the data has elapsed," Mitchell was told.[46] Mitchell responded that he needed more time, perhaps four more days, to complete the work.[47] The letter for more time went to Rockenbach, who had initiated the request for the officers' return in the first place. He then passed the letter to the adjutant general, who turned down the request and ordered the three returned to their jobs in the Military District.

Also on the morning of October 7, 1925, the army informed Mitchell that court-martial proceedings against him were to begin very soon. Mitchell had been scheduled to testify to the navy board investigating the crash of the *Shenandoah,* since Mitchell had had so much to say about it publicly. In a letter to Captain Paul Foley, navy judge advocate for the court of inquiry, Mitchell asked to have his testimony delayed on advice of his civilian and military defense counsel because of his pending trial.[48] Foley agreed to a delay. Ten days later a courier arrived from the Military District of Washington and gave Mitchell an order from Rockenbach that read, "You are advised that you are hereby placed in arrest. Your limits will be the City of Washington."[49] Apparently the army was not too concerned that Mitchell would flee the jurisdiction of the military court and show up on an island in the Pacific. However, the arrest order would keep him from traveling outside of Washington for speaking engagements, and there were indeed many requests for Mitchell to speak. Of some irritation was the fact that he could not, without the permission of Rockenbach, travel to Boxwood to see Betty and the baby. While Rockenbach would grant permission when Mitchell later requested it, the situation still remained that the army had control over his movements and a limit on his appearances.

Of greater distress to Mitchell and his defense team was the list of the members of the court. The man in charge was the senior officer on the active list of the United States Army, commanding general of II Corps, former commander of the 1st Infantry Division in the AEF, and former commander of the Hawaiian Islands, Major General Charles Pelot Summerall. It was widely believed that when John Hines stepped down as chief of staff, Summerall would take his place. Slender and athletic-looking, hair now turning slightly gray at the temples, immaculately uniformed, Summerall was known by many as the finest soldier in the army now that his mentor John J. Pershing had retired. No one despised disorder and indiscipline more than this West Pointer from Florida. This was the Summerall that Mitchell had blindsided in Hawaii in 1923, and this was the general officer that Mitchell had criticized in his report. If ever there was a head lion in the lions' den awaiting a miscreant Daniel, it was Charles Pelot Summerall.

NOTES

1. "Proposal to Create a Department of Aeronautics," *Congressional Digest* 4 (April 1925), 232.

2. *New York Times*, 14 December 1924, 24.

3. Burke Davis, *The Billy Mitchell Affair* (New York: Random House, 1967), 200–201.

4. Mason Patrick, "Cost of Our Wartime Aircraft," *Current History Magazine* (February 1923), 783–785.

5. *New York Times*, 7 January 1925, 21.

6. Memorandum from the Office of the Adjutant General to Mason Patrick, at the Direction of Secretary Weeks, 29 January 1925, in the William Mitchell Papers, Library of Congress, Washington, DC, Carton 11. (Hereafter cited as the Mitchell Papers.)

7. Ibid.

8. Ibid.

9. *New York Times*, 4 February 1925, 3.

10. Hugh Drum's Prepared Statement, January-February 1925, in the Hugh A. Drum Papers, in the Military History Institute Archives, Carlisle Barracks, PA, Carton 22. (Hereafter cited as the Drum Papers.)

11. *New York Times*, 5 February 1925, 1.

12. William F. Trimble, *Admiral William A. Moffett: Architect of Naval Aviation* (Washington: Smithsonian Institution Press, 1994), 149–151.

13. *New York Times*, 6 February 1925, 1.

14. Ibid., 7 February 1925, 2.

15. Note from Davis to Colonel Sherman Moreland, ca. November 1925, in Records Group 153, Records of the Judge Advocate General, General Courts-Martial, William Mitchell Case Records, 1925, Entry 40, Case Number 168771, National Archives, College Park, MD, Carton 9214-6.

16. *New York Times*, 6 February 1925, 1.

17. Memorandum from the Adjutant General to Patrick, 7 February 1925, Mitchell Papers, Carton 11.

18. Ibid., 11 February 1925.

19. Enclosure by Mitchell, 16 February 1925, Mitchell Papers, Carton 11.

20. Enclosure by Patrick, 18 February 1925, Mitchell Papers, Carton 11.

21. Rickenbacker to Mitchell, 19 February 1925, in the Edward Rickenbacker Papers, Library of Congress, Washington, DC, Carton 20. (Hereafter cited as the Rickenbacker Papers.)

22. Mitchell to Rickenbacker, 19 February 1925, Rickenbacker Papers, Carton 20.

23. Weeks to Coolidge, 4 March 1925, Mitchell Papers, Carton 11.

24. *Old Homes and Families of Fauquier County Virginia,* from W.P.A. Records, 351–352. Courtesy of the Fauquier Heritage Society, Warrenton, VA.

25. Historic House Files, Thomas Balch Library, Leesburg, VA.

26. From the Drum Papers, Carton 22.

27. Wheeler to Mitchell, 14 September 1925, Mitchell Papers, Carton 11.

28. Davis, *Mitchell Affair,* 210–211.

29. William Mitchell, "Some Considerations Regarding a Limitations of Armaments," *Annals of the American Academy of Political and Social Sciences* 120 (July 1925), 87–89.

30. Trimble, *Moffett,* 160.

31. *New York Times*, 28 July 1925, 13.

32. Captain J. Green's debriefing of General Ira Eaker, 1972, in the Ira Eaker Papers, Military History Institute Archives, Carlisle Barracks, PA.

33. Hicks to Mitchell, 11 September 1925, Mitchell Papers, Carton 11.

34. Mitchell to Hicks, 11 September 1925, Mitchell Papers, Carton 11.

35. S. C. Hodgkin to Mitchell, 20 May 1925, Mitchell Papers, Carton 11.

36. Charge Sheet, 24 September 1924, Mitchell Papers, Carton 38.

37. Davis, *Mitchell Affair,* 228.

38. Typescript Copy of Patrick's Testimony, 21–22 September 1925, in the Benjamin Foulois Papers, Library of Congress, Washington, DC, Carton 36.

39. Extract of Hines's Testimony, Mitchell Papers, Carton 37. Also see "Verbatim Report of Morrow Commission of Inquiry," *Army and Navy Journal* (26 September 1925), 1–24.

40. *New York Times,* 22 September 1925, 1.

41. Ibid.

42. Ibid., 18 September 1925, 1.

43. Ibid., 27 September 1925, Sec. VII, 5.

44. Mitchell to Rickenbacker, 1 October 1925, Rickenbacker Papers, Carton 20.

45. *New York Times,* 5 October 1925, 2.

46. Mitchell to Rockenbach, 7 October 1925, Mitchell Papers, Carton 11.

47. Mitchell to the Adjutant General, 7 October 1925, Mitchell Papers, Carton 11.

48. Mitchell to Foley, 7 October 1925, Mitchell Papers, Carton 11.

49. Memorandum to Mitchell from Rockenbach, 27 October 1925, Mitchell Papers, Carton 11.

TEN

DAMNED ROT

A MERICANS HAVE ALWAYS BEEN FASCINATED WITH SPEC-
tacular court trials—flashy defendants, brilliant defenders, tenacious
prosecutors, points and counterpoints. The United States of the 1920s had
its fair share of these—the Leopold and Loeb murder trial in Chicago in
1924 had the drama of wealthy defendants, hints of deep psychological
problems, and a struggle over the death penalty. The Scopes "Monkey" trial
in the summer of 1925 featured the brilliant mind of Clarence Darrow pit-
ted against the great orator, religious fundamentalist, and perennial presidential candidate William Jennings Bryan. The Mitchell court-martial
promised to be the type of courtroom drama the 1920s had come to expect
and relish. Here was a case that might well reach into the War Department,
possibly even to the White House. Also, unlike Leopold and Loeb or John
Scopes, Billy Mitchell was a well-known public figure, and the generals
who would judge the case were among the finest who had served in the
AEF during the Great War. Defendant and judges would have chests splat-
tered with colored ribbons, shining eagles, and stars. There was no murder
victim, no bespectacled school teacher, but there was Mitchell in beautiful-
ly tailored uniform, with his attractive wife and anxious sister in the courtroom facing stern-faced veterans of some of the bloodiest fighting on the
Western Front.

No face could have been more somber than that of Charles Pelot
Summerall, and his colleagues were no less filled with gravitas. Sitting
next to Summerall was Major General Robert L. Howze, a West Pointer
from Texas who had had extensive divisional command in the war; Major
General Fred Sladen, also an academy graduate who had commanded an
infantry brigade in some of the worst battles of the war; and Major General
Sidney S. Graves, another Texan with an academy ring, who had the unusu-

187

al experience of having commanded U.S. combat troops in Siberia. Born in Confederate Alabama in 1863 was Major General Benjamin Poore, an infantryman who led the 7th Infantry Brigade of the 4th Division into the St. Mihiel salient and into the rain and cold of the Meuse-Argonne. The remaining major general sitting on the court was a familiar face—Douglas MacArthur, the same MacArthur whose recommendation to West Point came from Senator John Lendrum Mitchell. Douglas MacArthur would forever be associated with the hard-fighting 42nd, or Rainbow, Division, and he had been a maverick reformer at West Point after the end of the Great War.

There were six brigadier generals on the court, five of them West Point graduates—all of them combat veterans of the war. The three prosecution lawyers, Colonels Blanton Winship, Sherman Moreland, and Lieutenant Colonel Joseph I. McMullin, came to the army from excellent civilian universities and law schools. Blanton Winship was designated as the Law Member of the Court, and it would be his job to interpret the manual for military courts-martial. No better choice could have been made; Winship had a fine reputation from his service with the AEF and was touted to be in line to become judge advocate general of the army, a post he would eventually fill. It was obvious that the fighting generals on the court could not know the ins and outs of military law, and neither could civilian counsel Reid. It would be Winship's task to fairly and impartially rule on motions and objections and to advise the prosecution, the defense, and the court as to the law prescribed in the manual. Winship's handling of his role during the court-martial fully justified his selection, and it was noted by all there, from generals to journalists, that he gave to the lengthy and rancorous proceedings the air of a fair trial. The court displayed a glittering array of tested officers, and when Frank Reid read the list he scribbled in the margin of the paper that this could well be a jury chosen to prejudice public opinion against Mitchell.[1] Much had been made of the fact that not one of the generals of the jury was a flying officer, but since this was a general court-martial, finding an aviator of proper rank would have been either an impossibility or a detriment to Mitchell's case. Certainly Mason Patrick, as chief of the Air Service, could not have been assigned, nor could Mitchell's replacement, Brigadier General James Fetchet. Benjamin Foulois, another ranking aviator, had been a bitter foe of Mitchell since the war.

The army assigned Colonel Herbert A. White, judge advocate of the VIII Corps at Fort Sam Houston, Texas, as Mitchell's military defense counsel.[2] White was a West Pointer from Iowa who had graduated from Columbia University Law School in 1898. He had seen service in the AEF and held the Distinguished Service Medal. While the teams of lawyers for the prosecution and the defense were taking shape, Colonel Moreland, a meticulous trial advocate, asked the army for an experienced Air Service officer to be assigned as a technical adviser and expert. Because Moreland believed that Reid and Mitchell would try to make national defense and air policy a cornerstone of the defense strategy, the prosecution would need to know quickly what Mitchell's lawyers were driving at. It was also imperative that whoever the army chief of staff assigned would not be one of Mitchell's devoted disciples. Before the trial began Major B. Q. Jones of the Air Service, a flying officer and veteran of the AEF, was detailed as the technical adviser, and his insights would be invaluable as the case went forward.[3] A general court-martial is the most serious of all military courts, and the army had put together a cast of characters that befitted the situation.

The Mitchell defense team pondered strategy to combat what seemed to be a pretty clear case, since Mitchell had informed the VIII Corps adjutant general that he had indeed made the statements attributed to him. What was to be decided, then, was whether Mitchell had a right to say what he did and whether it was correct. The next order of business, with the trial only eight days away, was to look at the jury and see if any of the general officers, especially any of the major generals, should be challenged from the jury. Mitchell and Reid focused on the president of the court, Charles P. Summerall, but challenging him would be tricky indeed given the respect he had throughout the army. Mitchell and Reid had to come up with a very solid reason for challenging Summerall, and they decided to base the challenge on Mitchell's inspection and report to Mason Patrick in 1923 about the Air Service in Hawaii. In making this challenge Mitchell was clearly attacking the competence of a senior officer who was touted to be the next chief of staff of the army. Mitchell went to work on a statement giving justification for the removal of Summerall from the jury, and he wrote several drafts before he and Reid agreed on the text. Mitchell had been in uniform long enough to know that if he went after Summerall and won, even if he were found not guilty, his days would be numbered as a soldier—but Billy Mitchell probably knew they were, anyway.

The room in which the trial would take place was in the old Emory Building, which once housed the U.S. Census Bureau. It was selected by the army because it was not too near to the War Department, and, frankly, Secretary Davis wanted a small courtroom to limit the crowds. He knew what massive crowds the Leopold and Loeb trial attracted in Chicago, and also what sort of circus the trial of John Scopes in Dayton, Tennessee, had been. It was serious business, the Secretary of War rightly stated when asked why the army had not hired a large hall for spectators. It would take several days to properly clean the courtroom, but it would be ready for very solemn proceedings by October 28, the day the trial was to commence. As nature would have it, after a very warm autumn in the East a raging storm struck that day with gale-force winds, rain, hail, and then a wet snow. Temperatures in Washington dropped a surprising thirty to forty degrees in the space of a few hours, and although there was some heat in the building the room would never be really warm as winter set in with a vengeance.

The day before the trial began, Mitchell's replacement, Brigadier General James E. Fetchet, gave a speech before the national convention of Sigma Chi fraternity at the Waldorf-Astoria Hotel in New York City. Fetchet said that Mitchell was indeed guilty of insubordination, but that was probably the only way he thought he could get his message before the public. "We of the Air Service want [the Air Service] run by airmen. The infantry and the cavalry are worthy men but they know nothing about the air," proclaimed Fetchet, who went on to say that the air had to be a separate service with its own promotion list and funds.[4] Fetchet would not have spoken as he did had he not discussed it with Mason Patrick before going to New York. The trial was quickly becoming less about Mitchell's violation of the Articles of War and more a round in the debate over a separate air service and the national defense.

The first day of the trial was a tedious one. Mitchell arrived in his best uniform and sporting a new bamboo cane, and his wife and sister were encased in heavy coats to ward off the bitter wet weather. Reid began a lengthy discussion of the propriety of the trial: did the president of the United States have the authority to order proceedings, and did the army actually violate Mitchell's right of free opinion and speech when it brought charges against him? A *New York Times* reporter wrote, obviously disappointed, that the fireworks were missing. The one thing that made reporters take notice was the mention of Calvin Coolidge's directive to discipline Mitchell. Would the defense have the nerve or indeed the right to call a sitting president of the United States to the witness stand?

Reporters who wanted courtroom drama got their wish when the jury was seated. Colonel Moreland asked the assembled officers if they had any prejudices in the case, to which they all answered no. He then, as required by regulation, asked if any would benefit by military promotion if Mitchell were found guilty, and again all responded no. Then Moreland asked if Reid had any challenges, and Reid challenged Brigadier General Albert J. Bowley, commanding general of Fort Bragg, North Carolina. The basis for this request for removal was a recent speech Bowley had given to a convention of the American Legion in Greenville, South Carolina, in which he stated that the infantry was still the backbone of the United States Army and that Legionnaires must not be misled by those who would say otherwise. Bowley, who had been an artilleryman in the AEF, said that he did indeed make that speech, but that he could be fair. The defense thought otherwise, and Bowley was excused.[5]

Then Reid dropped a bombshell in the middle of the courtroom by challenging Charles Summerall on the basis of hostility, prejudice, and bias against Mitchell personally. The court was shocked, and the general officers simply looked at Reid in disbelief, so surprising was it that anyone could possibly think that the senior officer in the Regular Army would be anything but a paragon of military justice. Mitchell and Reid had worked on their position for almost a week, with Mitchell preparing the basic challenge. Summerall had testified before the Morrow Board that he believed the Air Service was in good shape, and that it was his observation that morale among its members was good. The general had made a speech in New York City in which he castigated the "extravagant claims for a separate Air Service," a fact he did not dispute, but in his address Summerall did not cite Billy Mitchell by name. Reid then went on to the report Mitchell had filed in December 1923 as the main source for the suspected hostility. Mitchell's personally prepared document said "that the Air Force under him [Summerall in Hawaii], and the whole system of defense was inefficiently handled, badly organized, and the ignorance of its application was manifested by himself and his staff." In the final draft of this statement Mitchell added, "This handling and administration of these defenses would lead to certain defeat in case of war."[6]

Summerall's eyes were full of fire, and as Reid spoke his face became flushed with anger. Mitchell's statement then cited the antagonism between Summerall and the commander of naval forces in Hawaii, which was so bad they would not attend social functions together. He inserted Summerall's letters to Mason Patrick claiming that remedial steps in regard to the Air Service in Hawaii were underway, but according to Mitchell's statement all Summerall did was to sidestep the critical issues contained in his report. Billy Mitchell's solution to all of this was a unified air commander for Hawaii who would be above, or at least away from, these clashes of egos and personalities. Trying to be as composed as possible under the circumstances, Summerall told the court that until that moment he had no idea what bitter feelings Mitchell had for him, and that under no circumstances could he remain on the court. He was then excused, and a very angry General Howze took over as senior officer. Major General Fred Sladen was challenged with no cause given, and it was obvious to all that he was very happy to vacate his seat, grab his hat and coat, and follow Summerall out of the building.[7]

On October 30 Mitchell responded, in a loud voice, "not guilty" to all of the charges made against him, and Frank Reid indicated that he intended to call President Coolidge and Secretary Davis to testify. This caused a great deal of squabbling throughout Friday, October 31. By the end of the day it was agreed that a subpoena for President Coolidge would be out of order, but the status of Secretary of War Davis was not too clear.[8] By the end of that first Friday, Mitchell and his wife found themselves in a quandary. Betty Mitchell had left the baby in Detroit in the care of her parents while she was in Washington giving moral support to her husband, and she had planned to return to her parents' home each weekend to see the child and tend to business there. Mitchell had intended to go with her, but the particulars of his arrest stipulated that he could not leave the environs of Washington.[9] She would have to go alone, with Mitchell remaining in Washington to work with his defense team for the next week's proceedings.

Frank Reid might not have been the best defense counsel for a military court, but he was detailed in preparation and tried to keep Mitchell on track as the trial went into the second week. He refused any payment for his work, but the costs of maintaining large apartments and bringing witnesses to Washington, not to mention stenographic help, were mounting. Elizabeth Mitchell's father, a lawyer for the Packard Motor Company, had to bear the cost of the trial because his son-in-law could not afford the bills, which were growing every day. It was pretty certain that the court would present its case rather quickly, since Mitchell did not deny making the statement or giving it to the reporters in San Antonio in a lengthy document. It would fall to the defense team to show that what Mitchell said was right and that he acted in the best interests of the Air Service, and therefore in the best interests of the army and the nation. Mitchell and Reid sat down with the statement and went through it paragraph by paragraph, line by line, and designated sixteen major points they wished to make about the condition of aviation.

Point number four, for example, concentrated on the administration of air matters and the damage it caused to morale. In the margin of his notes Mitchell wrote that he planned to call as witnesses Thomas D. Milling, Herbert Dargue, and several other aviation officers. Point number fourteen dealt with the army's anti-aircraft experiments, and he penned in the names of Brigadier General Johnson Hagood, who had just blasted the army over what he perceived to be failures in that area, and Major General Mason Patrick, who, like Mitchell, had little faith in the capabilities of the army's anti-aircraft system. Hagood's testimony would be important because during the war he had been instrumental in creating the staff system for John J. Pershing. Just recently Hagood had called for creating a separate Air Service, as the army had done with coastal artillery.[10] Despite popular views of the Mitchell trial, there were definite points Mitchell and Reid wanted to make to bolster the plea of not guilty. It would not be a haphazard calling of witnesses, and there would be many to be called before the court.[11]

Once Mitchell had finished with his statement, Lieutenant Clayton Bissell, who had known Mitchell for several years, and William H. Webb, a defense-minded lawyer from the House of Representatives, went over Mitchell's points with great care. Webb had been personally selected by Reid for his in-depth legal knowledge, and Bissell understood the nuances of military language.[12] Because the prosecution stated that it would call only seven witnesses—six Texas newspaper reporters and Colonel George K. Hicks, adjutant general, VIII Corps—it was clear that the only possible defense would be proving Mitchell's statements to be true and not prejudicial to the order and discipline of the U.S. Army. To do that, Reid said to the press, he had to be allowed to go beyond Mitchell's act of issuing a statement and a document—he would have to call witnesses to the stand.[13] It was a stretch, especially with the court already in shock over the removal of Summerall, but it was the best that could be done. To really complicate matters, some of the junior officers who were to testify had commanding officers sitting on the court.

Colonel Moreland and Lieutenant Colonel Joseph McMullin realized very quickly that this trial would go way beyond the original charges and would involve a defense that centered on the veracity of Mitchell's statements. Along with Major Jones, they combed Reid's opening statement for points they could make when Moreland presented the prosecution's opening arguments. They were detailed in their work; in fact, much more detailed than were Reid and Colonel H. A. White. The defense was hindered by Mitchell's attitude toward the proceedings and his air of unconcern, which angered Reid and White. The prosecution spent long hours after court going over the transcript of the day's work. In their preparation to answer Reid's opening statement the lawyers identified ninety-one points that Reid had made in defense of Mitchell's actions. Major Jones submitted a list of points the prosecution could respond to, but the prosecution refined the points of rebuttal to a mere fourteen and suggested that at least two be dealt with by the Navy Department when its representatives were called to testify for the prosecution. Moreland and his team put in long hours of work focusing on the goal of a guilty verdict.[14] Given the obvious fact that the Mitchell defense would hit at any possible failure in the area of airpower, Moreland relied on a great deal of support from the War Department, the Navy Department, and the General Staff. The latter had already told Moreland that their main spokesman on the witness stand would be Brigadier General Hugh Drum, and Drum was salivating over the chance to hit at Billy Mitchell. Moreland also requested that the judge advocate general assign two more trial lawyers to his team because the workload had grown in proportion to the wide-ranging nature of the defense. The judge advocate general obliged the hard-pressed Moreland with two lawyers, one of whom was Major Allen Gullion, known as a merciless and very effective cross-examiner.

A number of newspaper reporters viewed the trial of Billy Mitchell as a vehicle for getting the public and the politicians to hear the Mitchell message of air defense and a unified air force. Not every editorial writer was enamored with Mitchell and his methods, but some newspapers, especially the Hearst chain, were ready to do anything to support Mitchell and to sell copies. In early October there appeared in the *New York Evening Journal* an editorial entitled "A Separate Air Force," which said, "Colonel Mitchell, who has the courage to tell the truth, and whom politicians, much as they would like to, will not have the courage to dismiss, tells exactly what the nation ought to have." Not content with support, the newspaper included a message that the reader could cut out and send directly to President Coolidge. The coupon stated, "I have read the discussions about a separate air force. I agree with Colonel Mitchell. I hope that you will disappoint the politicians and lobbyists that want to use the navy for profit, selling the government obsolete battleships, and otherwise urge the establishment of a separate air force and advise putting Colonel Mitchell at the head of it." Thousands of the 2-inch-by-4-inch coupons poured into the White House, many with letters attached. The response was so great that the coupons could not be processed by Coolidge's staff, and they were sent to the chief clerk of the War Department. Petitions and personal letters also came into Washington by the thousands.[15]

No bureaucracy is immune from political pressures, and the War Department was certainly no exception. It made sense to Frank Reid, who was a politician, to push Mitchell's defense based on the correctness of the charges, since Mitchell's popularity was at its zenith. It was certainly no accident that when Mitchell had his photograph taken by eager newspapermen he was with his attractive wife, often holding Lucy, his infant daughter. There was deep concern in the War Department that Mitchell might very well wind up as a cabinet member with control over the air. Army Chief of Staff John Leonard Hines tried to shun publicity as much as possible, and he made every effort to keep his immediate staff, except for Hugh A. Drum, out of the limelight. The ambitious and highly articulate Drum was designated to present the army's position at the trial when called. Of course Mason Patrick, as head of the Air Service, could not remain above the battle. As Mitchell's fight over the Air Service reached its dramatic conclusion there was sober reflection that was quite different from the Hearst coupon campaign. *The Independent* stated, "There is more than an even chance that Colonel Mitchell will overplay his part and lose his audience. The public sympathizes with a victim of official tyranny, but it quickly tires of the individual who keeps on shouting that he alone is right and that everyone else is wrong. . . . Colonel Mitchell, unless he controls his pen and his tongue, will be impatiently dismissed by the public as a know-it-all. He protests too much."[16]

While the American Legion was wildly enthusiastic in its support for Mitchell, egged on by men like Eddie Rickenbacker, other veterans' groups were not. Retired Major General Mark L. Hersey, president-elect of the Military Order of the World War, was one of Mitchell's critics. He told John L. Hines that he was ready to carry in the order's monthly bulletin any statement the army chief of staff might care to make on the Mitchell case.[17] Hines responded that although he had no new statement, he would be happy to have his testimony before the Morrow Board reprinted in part or in its complete form. He then told Hersey that there were several admirals who would be willing to contribute their thoughts as well.[18] That was an understatement—because there were many naval officers, including those with aviation experience, who were ready to go after Billy Mitchell. It was not a question of having expert and appealing witnesses in navy blue to sit in the witness chair. It was a matter of which were the best to select.

When the trial resumed on Monday, November 2, 1925, the courtroom was crowded and even more uncomfortable, as spectators with heavy, damp coats pushed into the room. When General Howze had dismissed the court for the weekend it was thought that the case could be disposed of by Wednesday or Thursday of the next week, but that was quickly seen to be in error as Reid presented his witness list. Seventy-three defense witnesses were identified on Reid's list, which also contained a request for massive documentation pertaining to bombing results (much of which was classified as confidential), materials presented to the Morrow Board, and other documents from the War Department, Navy Department, and the Agriculture Department. President Coolidge's personal secretary was on the list, and Reid wanted him to personally present the Morrow Board documents.

Despite the exhaustive defense witness list, Howze got the court-martial moving and called several San Antonio newspapermen who testified to what everyone agreed to: Mitchell had indeed made statements and given them a carbon copy of a longer statement. Colonel Hicks from the VIII Corps was obviously not happy with his role as a witness, and when Reid asked him if Billy Mitchell's statements had caused breaches of discipline at Fort Sam Houston, he replied simply: no. Colonel Moreland was on his feet, objecting to the question, but Howze overruled him. As a good defense counsel would, Reid asked the question again, and Hicks responded, "In my judgement they did not."[19] Reid had every reason to be pleased with the first day of real testimony.

On Tuesday, November 3, the day started out well again for Reid and Mitchell when Howze ruled that the defense was well within its rights to call their witness list and ask for documents bearing on the case. Reid then asked for a dismissal of charges, knowing that Moreland would object and Howze would refuse, which he did. Howze was furious at the slow pace of the trial, and he ordered Reid and the defense team to meet with the prosecution and agree on who was to be called and how long they would be on the stand. The Texas cavalryman wanted the trial moved along and made no bones about it. He then dismissed the court and glared at the lawyers as they filed from the room to their meeting.[20] The next day Reid presented another witness list to Howze, which omitted cabinet secretaries and Coolidge's personal secretary. The document list, however, was intact, much to Colonel Moreland's distress because, as he said, the assembling of so much paper could take a lot of time, and he did not want to see the entire controversy over airpower dragged into the court-martial. But that was exactly what Reid and Mitchell wanted to happen.

The Mitchell trial actually fell into three phases: first, the seating (or unseating) of the court; second, the preparation of prosecution and defense motions and lists; and third, the actual presentation of the case and the defense. After the newspaper reporters and the uncomfortable Colonel Hicks testified, Moreland called Henry S. Parsons, chief of the Periodicals Division of the Library of Congress, to the stand. Reid and Mitchell were puzzled as to why the scholarly Mr. Parsons would be a witness, but they realized quickly that his testimony would be damaging. Parsons established that Mitchell's statements were indeed carried in all of the major newspapers by producing copies of the *New York Times, St. Louis Globe-Democrat, Chicago Tribune,* and *New Orleans Times Picayune.* Parsons stated that Mitchell's statement was carried in toto by all the major newspapers and was therefore read by the vast majority of literate Americans, but Reid saw an opening for the defense. If the statement was read by so many across the country, it stood to reason that the veracity of the statement was key to Mitchell's claim that he did nothing wrong if indeed he stated the truth. Moreland had opened the door a crack by calling Parsons as an expert witness from such a highly respected institution as the Library of Congress, and Reid threw it wide open.

When the trial began again on November 10, it was pretty clear that the

Mitchell defense team was ready to discuss the whole air controversy. Major Carl Spatz testified, but he did not say anything really new. Rather, his testimony recounted the same arguments Mitchell and his airpower disciples had been giving since 1919. The main thrust of the defense went beyond those in the courtroom who were familiar with Mitchell's career. Reid was aiming at the public, which was not as well informed as the officers sitting in the court. The army would have to adjust its strategy, because the defense was not contesting at all the issuance of the statement following the crash of the airship *Shenandoah*. It would now have to go after Mitchell himself as an officer out of control. After Spatz and other witnesses finished, Reid surprised the court by announcing that he intended to call Mrs. Margret Ross Landsdowne to the stand to testify that she had been pressured by the navy to give false testimony to the board investigating the crash of the airship that took her husband's life. Colonel Moreland immediately protested that the board was a navy matter, but Howze, acting on advice from Blanton Winship, overruled him because it was the crash of Landsdowne's airship that had brought about this whole trial.

The Department of the Navy had good reason to be happy with the predicament of the army in trying one of its own, but with its own investigation of the *Shenandoah* crash not going well and with Howze willing to explore the crash, there was a growing sense of urgency to control adverse public reaction. This was further strengthened when Colonel White sent to Moreland a two-page request for Navy Department documents ranging from verification of the cost and the construction of the *Shenandoah* to reports on the bombing trials in 1921, to projected budget requests up to the fiscal year of 1927.[21] The navy assigned two additional persons to its Press Relations Section, and it was clear that their role was damage control. Lieutenant Richard W. Gruelick and Helene M. Philibert were given the difficult tasks of monitoring the trial and dealing with information that might adversely impact the navy. Philibert was an experienced civilian employee, having been in the Press Relations Section for years. That particular section was part of Naval Intelligence and had to handle all manner of sensitive issues.[22] With Margret Landsdowne prepared to sit on the witness stand on behalf of Billy Mitchell, and with the whole *Shenandoah* question opened for consideration, the Navy Department was ready to begin a public relations campaign of its own.

While the court prepared for Margret Landsdowne's potentially explosive testimony, the defense had to puzzle over what to do with classified documents that Mitchell's friends were willing to make available. At the same time Mitchell was informed that his first wife, Caroline Stoddard Mitchell, was back in court in Wisconsin seeking an increase in child support payments. Handling that issue would be just as complicated as dealing with classified material, because Mitchell was winning points with the press, if not the jury, and this reminder of a personal scandal could damage the public image of Mitchell as a selfless crusader, an image Reid was pushing hard with the press. Unable to leave Washington, Mitchell would have to deal with that problem through Milwaukee lawyers; he certainly did not want to petition Howze for a delay and an exemption to his arrest order. Mitchell had been slow at times in making the original court-mandated support payments and seemed uninterested in his three children, manifesting the same sort of indifference that his father had with the children of his first marriage. Now Caroline was asking for personal alimony of $2,000 and an increase in child support to $4,800 per year. Mitchell's lawyers pointed out that Billy's army pay plus his income of about $4,700 a year from his mother's estate would not cover the amount Caroline asked.[23] Mitchell chose not to disclose that most, if not all, of the funds to buy Boxwood and maintain his horses and thoroughbred dogs came from Betty Mitchell. Instead, he argued that Caroline's personal wealth at that time was almost $200,000, a much greater sum than he himself possessed. The court did adjust the payment demands upwards, though not granting all that Caroline had demanded. This was a distraction that Mitchell really did not need as the court-martial progressed.

The question of whether to produce the requested classified material was still to be answered. Billy Mitchell had said again and again that the nation needed a cabinet-level secretary to oversee and coordinate all matters, military and commercial. His argument focused on the confusion and chaos in such a new industry. He was handed a golden opportunity to drive the point home when First Lieutenant Corley P. McDarment of the Air Service's Information Division sent a document to Reid and to Mitchell. What made McDarment's memorandum so explosive was that he was at that time the custodian of secret air observation photographs. In 1924 he wanted a number of aerial photographs of U.S. military installations downgraded in classification from "secret" to "official use only" because commercial aircraft could photograph those installations anytime they flew near them. He sent his recommendation through the chain of command, and the document was sent to various agencies in the War Department. His request was refused, but he learned that the Coastal Artillery wanted authority to fire on any aircraft, commercial or military, that flew near a military installation! "The Coast Artillery would therefore shoot down airplanes that lost their course, or were driven over forts by storms, etc. This shows a lack of knowledge of flying," McDarment wrote. He also included a secret document outlining the various steps taken to arrive at the decision to do nothing as far as aerial photographing was concerned.[24] This was dangerous business for the lieutenant, who had worked with Mitchell before, because he had sent a copy of a secret document to a defense team with civilian lawyers. Reid was wise enough to know that the introduction of this material would not help the trial and could cause Mitchell a good deal of harm.

McDarment's memorandum came at a time when Reid and Mitchell were preparing for the testimony to be given by Margret Landsdowne. One might wonder why the defense did not use at this point such potentially damaging information pointing to a lack of knowledge and a lack of coordination in air matters. It would have been very easy to point to the folly of having no central air authority when a branch of the army wanted authority to fire upon and shoot down commercial aircraft. For good reasons this whole line of defense was best not brought out in court. Besides, as Reid pointed out, Margret Landsdowne was worth more than a boring, tedious debate over anti-aircraft policy. Mrs. Landsdowne had been testifying before the navy's board of inquiry about the crash of the *Shenandoah,* and she had stated that the navy had urged her to give false testimony. Reid had already prepared the press for her statements before the court-martial by telling them that she had been asked to alter her story about her husband's concerns over taking the *Shenandoah* across mountains into Ohio, where information about the weather had been lacking.[25]

Margret Landsdowne's presence in the courtroom was bound to cause trouble for the prosecution simply because she was young, very attractive, soft-spoken, dressed in widow's black, and speaking about her late husband, a respected commander of what had been the pride of the navy's fleet. Following the standard questions about identity and so forth, Reid moved quickly to the crux of her testimony. He asked her if Captain Paul Foley, judge advocate for the *Shenandoah* inquiry, had visited her with an unsigned statement that she believed contained untrue statements about her husband and the fatal flight of the *Shenandoah.* She stated that this was true and added, while looking directly at Howze, that the Foley document indicated that Landsdowne, as a naval officer, was fully and willingly prepared to take the ship out as he had been ordered. In fact, Mrs. Landsdowne stated, the final mission of the *Shenandoah* was a political one, not strictly military in nature. Mrs. Landsdowne completed her testimony and left the courtroom after a few carefully chosen and respectfully asked questions by Moreland.[26]

The Landsdowne testimony was the stuff of which movies were made — high drama, a beauty ready for an onslaught by high-ranking officers protecting some sinister organizations, the grieving but gallant widow standing for truth. The newspapers loved it, and many editorial writers called for charges against Captain Paul Foley and for the resignation of Navy Secretary Wilbur.[27] For Mitchell it was a moment of personal triumph, since he had claimed when the *Shenandoah* went down that the navy would try to cover up any wrongdoing. What were navy airships doing flying over mountains into the heartland of America anyway, Mitchell had challenged in September. Billy Mitchell and Frank Reid failed to see that this court of army officers was sure that the navy was perfidious. Most of the officers of the court were generals who had graduated from West Point, and they were used to muttering "Go Army, Beat Navy!" But that did not mean that the lovely young widow's testimony added anything to Mitchell's overall defense. When the navy's *Shenandoah* inquiry board exonerated Paul Foley of any wrongdoing and subpoenaed Mrs. Landsdowne back to the inquiry without her civilian counsel, who was denied access to the court, the generals were sure that this was another example of "The Navy," the same fellows who needed a good thrashing on the football field in the late fall every year. How much good it would do Billy Mitchell was another question again.

With Margret Landsdowne's testimony the focus of the trial shifted toward the crash of the *Shenandoah* and the navy's role in the tragedy. Mitchell had in his possession a number of documents about the airship and naval policy that he began to study in depth. The day after Mrs. Landsdowne's testimony, Paul Foley asked to be relieved as the judge advocate and to be allowed to appear before the Mitchell court-martial. At the same time Colonel Sherman Moreland asked that Mrs. Landsdowne's testimony be stricken from the records, a request Howze denied. To irritate Howze even more, Reid and Moreland got into a lengthy tussle over the role of the General Staff in the trial. Reid pointed to Major Francis Wilby, who was seated at the prosecution table, and stated that he had been sent by General Dennis Nolan to give the Staff's guidance to the prosecution team. Then both men got into a long, drawn-out argument over documents requested by both sides. The mass of paper, Colonel Moreland argued, could necessitate a recess of about two weeks. At this Howze, who could take no more of the lawyers' constant wrangling over what appeared to the Texas cavalryman like so many angels and so many pins, exploded with a loud and angry "No!" After a few witnesses and more confusion over documents—would the court see originals or copies?—Howze (acting on the advice of Blanton Winship) did grudgingly recess the court for the examination of documents, but only for two days.

General Howze had repeatedly warned both Reid and Moreland that he wanted the trial to move along. It was a very costly procedure, and it was drifting too far afield for his tastes. General Howze had no idea what a financial nightmare the court-martial of Billy Mitchell would become. As the trial he had thought would last only a few days dragged on, costs mounted. The prosecution, reacting to the list of defense witnesses, drew up a preliminary roster of their own that totaled eighty-eight military personnel and civilians. This meant that the government had to pay train fare from as far away as Texas and California, and there were three potential witnesses in the Hawaiian Islands. Civilians had to be put into hotels, compensated for their loss of work, and, like their military counterparts, given a per diem of $1.50. To try to offset the mounting bills, sworn depositions were taken from as many individuals as possible, especially those in California and Hawaii. It did not matter to Howze how much Billy Mitchell's father-in-law was spending on his defense, but he knew it was a great deal of money. The office of the judge advocate general had authorized Lieutenant Colonel Joseph I. McMullin to contract for court reporting services at what seemed to be a reasonable rate for a true copy and four carbon copies of the transcript: 25¢ per hundred words. William M. Day, an experienced local court reporter, took the job. By the end of the trial the transcript had reached almost 4,000 pages at over 50¢ per page, and that did not include mandated changes, deletions, and additions. At a time when the budgets for all of the services were slim indeed, and with no chance for supplemental appropriations from Congress, the court-martial was fast becoming a financial burden for the army. Howze was well aware of the alarm in the War Department over the cost of the trial, and he saw the possibility of the trial stretching into 1926 if the defense was allowed to pursue an unchecked attack on the navy. Blanton Winship was in agreement with Howze and, though he bent over backward to accommodate the defense, did not want Mitchell to put the navy on trial. Even so, it was the crash of the airship that had triggered that statement from Mitchell and brought on the proceedings. Allen Gullion prepared a long list of questions to ask navy personnel and at the same time a list of questions to ask Mitchell when he had an opportunity to cross-examine him.[28]

The defense wanted to approach the *Shenandoah* crash from three angles: first, the construction of the airship was faulty; second, the mission that led to the tragedy was unnecessary; and third, the navy did not provide any safety margin for Landsdowne and his crew. These were basically the charges Mitchell had leveled against the navy in his Fort Sam Houston statement and document. By just looking at the list of documents Colonel White had requested for the defense, one could see that Reid intended to put the navy on trial. Winship decided that because the navy was conducting its own board of inquiry it was useless to duplicate those efforts and possibly drag the court-martial into the new year. Very quickly into the testimony on the *Shenandoah* crash, Howze exploded and said that he had no intention of hearing over and over how the *Shenandoah* was constructed and maintained. Winship jumped in as the Law Member of the Court and ruled an end to the tedious and technical testimony.[29] This would not be a trial of the navy despite the defense's best efforts.

Reid was furious, and when the prosecution called Commander Kenneth Whiting of the navy to the stand he continually objected to any testimony. Whiting was an expert on the subject of aircraft carriers, and the prosecution wanted to show that Mitchell's writings and statements about carriers were faulty in the extreme. The proceedings again dissolved into wrangling between Reid and Gullion, and the court was treated to long discussions about the navy's policy over the construction and deployment of aircraft carriers. As Whiting's constantly interrupted testimony got into British aircraft carrier policy, the whole thrust of his comments became lost in a welter of objections, counter-statements, and rulings by Blanton Winship. Howze was becoming more and more angry with each confrontation between Reid and Gullion, and it was clear that the court and the spectators were lost in the torrent of words and segmented testimony.[30] Even the experienced reporters from the *New York Times* seemed to have difficulty making sense out of the proceedings.

The *Shenandoah* crash continued to occupy much more time than Howze wanted, and despite his efforts and the rulings by Blanton Winship there were more witnesses. Reid's cross-examination of each one became more detailed. One witness who damaged the defense's case was Captain G. S. Lincoln, the director of Ship Movements in the office of the chief of naval operations. Lincoln was the officer directly concerned with routing the airship through the Midwest, and his appearance on December 4 promised to be contentious at best. Lincoln produced a directive from the chief of naval operations, dated August 12, which gave guidance to Zachary Landsdowne. He pointed to paragraph three, which said that "Should the dictates of safety and the weather conditions existing make it advisable, the commanding officer of the *Shenandoah* is authorized to make modifications in the above itinerary as he deems necessary." The document stated in clear terms that the flight of the *Shenandoah* was to promote public interest in the navy and to fire the imagination of potential recruits for the service. In his cross-examination Reid tried to make much of this, but his questions pointed to a problem. The army, including the infantry and the Air Service, also had such public relations events, ranging from infantry weapons demonstrations to the Air Service being very much in evidence at air meets, county fairs, and the like. Mitchell himself had frequently been in attendance at such air competitions, especially the ones at Selfridge Field, Michigan. To the press and to the jury it seemed that Reid was engaging in "the pot calling the kettle black," with little understanding of legitimate, time proven recruiting tools. Reid introduced a telegram from Landsdowne to Lincoln sent on August 25 stating that he wanted to alter some times in the *Shenandoah* schedule. Lincoln said that he had received the telegram and that the chief of naval operations' guidance had not changed: as the commander on the spot, Landsdowne could alter times as long as he told the press so that the public could be informed of the changes. Reid continued his cross-examination but did very little to change the perception that the navy had taken safety and climactic conditions into account and allowed Landsdowne latitude in decisionmaking. Captain Lincoln's presence on the stand and his command of the facts cast a shadow over the testimony of Margret Ross Landsdowne, who had said nothing about the guidance from the navy about safety, weather concerns, and the commander's latitude in the routing schedule. Much had been gambled in putting Mrs. Landsdowne on the stand; and her demeanor as the pretty young grieving widow was effective, but with Lincoln's appearance for the prosecution much of that advantage was lost.

The officers of the court simply looked off into the voids of space as Reid and Moreland blithered. Douglas MacArthur spent much time looking around the room, some officers tried shifting to find comfort in their army-issued hard-bottomed chairs, and Howze looked like he was dreaming about the "old days" in Texas when he could have just hanged one or two lawyers and put an end to it all. Lost in the welter of objections and complaints and finger-pointing was Billy Mitchell. On November 15 several New York newspapers claimed that Margret Landsdowne had agreed to appear in a new Broadway play entitled *Just Beyond*. Besieged by reporters, the widow indignantly denied anything of the sort.[31] On November 16 Mitchell was able to regain the spotlight when he made a radio address that, to no one's surprise, attacked the battleship navy, the inability of the army to recognize airpower, and the need for a unified air service under a cabinet secretary.[32] The radio talk irritated the commander of the Military District of Washington, who had restricted Mitchell's movements to avoid just such an incident. Mitchell had been denied the opportunity to speak to the New York American Legion on Armistice Day, but someone overlooked a modern invention, the radio, and Mitchell had taken advantage of it.

If any group was enjoying the front-page spectacle that the Mitchell court-martial had become it was the Democratic Party. Former AEF Captain Samuel Taylor Moore, who had commanded a balloon company on the Western Front, was commissioned by *The Independent* to write a series of articles on the trial and what it meant. He understood, as did many observers, that the whole question of airpower and the Mitchell court-martial placed President Calvin Coolidge in a difficult situation. His legislative agenda dealing with tax reduction (a program dear to Silent Cal's Republican heart) and the World Court's relationship with the United States was in danger. Everything would be overshadowed by questions concerning airpower, questions the Democrats were willing to keep asking. Moore wrote that Mason Patrick, "the competent Chief of the Army Air Service has expressed himself as forcibly, more temperately, and therefore more intelligently than his former first assistant." The makeup of Congress, where the Republicans held a slim majority, could very well be altered in the 1926 elections, and a convicted (as Moore opined he would be) Mitchell with a severe sentence could be a spark in some districts to turn out Republicans and elect Democrats. Moore went on to observe that with congressional elections so close and so hotly contested, the Mitchell bloc in Congress would become noisy and would challenge Coolidge, his administration, and the service departments.[33] Moore and a number of objective reporters understood that there was a great deal at stake in the Mitchell trial, more than just Mitchell's ideas and his future. The National Democratic Committee made it very clear that air policy would indeed be an issue in the 1926 congressional races.

On Frank Reid's witness list for the upcoming days were Eddie Rickenbacker and Admiral William S. Sims. Sims appeared first and blasted navy policy, calling members of the Navy Department "ignorant and unfit." Sims, who had been a staunch advocate for airpower, stated that in his opinion the navy just "[bumped] along from day to day. As far as I can see no effort is being made to get a policy."[34] In the spring of 1925, in a debate in print sponsored by the *Congressional Digest*, Sims had written, "The most important thing for the defense of America is submarines and airplanes. . . . I think that air forces will play a most important part in the next war."[35] Before the Morrow Board the retired admiral and former head of the Naval War College had stated that battleships were indeed vulnerable to air attack and that Mitchell was correct in his assessment of the *Ostfriesland* bombings.[36] Rising to question Sims was Major Allen W. Gullion, the intense prosecutor, but despite sharp probing he was unable to shake Sims.

The Rickenbacker testimony was predictable but more interesting because of his status as the Ace of Aces in the Great War. In the almost four-thousand pages of the transcript of the trial Rickenbacker's appearance was fairly short. Reid's questions were mundane, focusing on the failure of anti-aircraft fire; a poor, slow supply of parachutes for aviators (balloonists had most of them); and the unsafe nature of American aircraft. Very carefully the prosecution cross-examined Rickenbacker. He was asked if it was not true that the German legend Baron Manfred von Richtofen (which the prosecutor pronounced as Rickover) had been brought down by anti-aircraft fire? Yes, Eddie Rickenbacker was aware of that. Then he was asked if he knew that the 23rd Anti-Aircraft Battery had brought down nine planes with about 6,000 shots fired, to which Rickenbacker replied, "Were those all German or some American?" That led to a rather dry discussion of altitude and bomber protection—not very fiery stuff.[37] When Eddie Rickenbacker left the stand he nodded and smiled at Mitchell and Betty, but the prosecution's handling of this testimony was solid. Rickenbacker was never given a chance by either Reid or the prosecution to focus directly on Mitchell's real defense, the truth of his statements.

After a few other witnesses Reid indicated that the parade of defense witnesses was over, and, while he felt they had proven the truth of Billy Mitchell's statements, he would call Mitchell himself to the stand on November 23. Reporters in the courtroom noticed that Reid's direct examination was careful and methodical. When Reid sat down and Major Gullion approached Mitchell the air was charged with electricity. Gullion had waited for this moment for weeks, and he had prepared by taking copious notes and by observing Billy Mitchell himself. Gullion had decided that his cross-examination would take two forms: attacking Mitchell's credibility as an expert on airpower and irritating Mitchell's ego to the point that he would answer with careless and antagonistic responses. During Reid's examination of Mitchell and from Mitchell's own articles, books, and testimony before boards and congressional inquiries it was clear that Mitchell wrote and spoke in generalities, with vast, sweeping phrases and concepts. Gullion would force specifics, showing, he was sure, that Billy Mitchell did not have command of the details and facts. That Mitchell was a well-educated and highly intelligent man Allen Gullion had no doubt, but he also felt that Mitchell's view of himself as a major force in the development of airpower and aeronautic policies was his Achilles' heel. If Gullion could puncture that facade of self-importance, Mitchell could be thrown off stride, forced to react to Gullion's relentless questions. He would be put on the defensive, like an injured boxer who could only try to ward of an opponent's well-aimed and well-timed blows. Gullion wanted to begin with as sharp an attack as possible to set the tone for the entire cross-examination, and he would use Billy Mitchell's love of the limelight to do it. There was no question that Mitchell relished his appearances before congressional committees, where he could lecture, in his style, on airpower and defenses issues. Would-be supporters on those committees would allow Billy Mitchell to say what he wanted, and the press would duly report it. What then was more important—the message or the messenger? Gullion decided that it was Mitchell the messenger who was speaking freely to the nation's lawmakers. There was no better place to start the cross-examination than with Mitchell's continual trips to Capitol Hill. Gullion would build slowly and then strike at Mitchell's long-proclaimed belief that the proponents of air defense had no fair hearing in the debates over American's defense policies.

The army General Staff, which had been highly irritated by many of Mitchell's charges, had prepared a number of detailed papers dealing with many of the allegations leveled by Mitchell. Of great importance to the staff, and to Brigadier General Hugh A. Drum in particular, was the assertion that officers had been muzzled by their superiors. Gullion would make great use of a position paper that traced army guidance on testifying before committees of the Congress.[38] Gullion personally held the view that Mitchell knew of this guidance, which allowed officers to testify freely so long as they did not openly discuss classified materials or engage in some sort of tirade. Gullion had, he believed, all of the high-caliber ammunition he needed to bring down Billy Mitchell.

There was no exchange of pleasantries as there had been with other witnesses; Gullion, known for his relentless cross-examination techniques, was ready to go to work. And go to work he did, boring in on Mitchell's statements. Reid rose several times to caution Mitchell to answer either yes or no, or give short and direct responses. As Gullion hit at Mitchell on various subjects from the costs of patrolling the Pacific with submarines to the results of bombing tests, Mitchell responded that what he said in his statements was a matter of opinion. "Since all of your statements were not facts, and were only based on your opinion," Gullion countered, "will you please state the source of your information so that we can differentiate between your opinions and your imagination?"[39]

Billy Mitchell had long ago opened the door for Gullion's devastating attack. In a letter to the editor of *The Independent* in April 1925, he wrote, "Every military airplane can be used in time of peace for some useful undertaking not necessarily connected with war. Every pilot employed in civil aviation can be used in case of war and is ninety percent efficient at least in time of peace. Every mechanic used in civil aviation is one hundred percent efficient in time of war."[40] Every combat aviator who read this type of Mitchell-speak had to recall the hours of training in flying in formation across the front and in weapons maintenance and gunnery, and old AEF mechanics probably wondered how much real training for combat a civilian mechanic had. It was this type of unsupportable generalization that Major Gullion could pounce on with a vengeance, and he did. Gullion was demanding specifics Mitchell simply did not have. The word *opinion* was heard continually from Mitchell.

"The whole case lies on the credibility and veracity of the defendant," Gullion told the court, and he was drawing blood. When asked about his statement in San Antonio, Mitchell replied, "My sources of information were studies I had made in the Northern hemisphere and what I have seen, looking into the future to see what our national defense should be."[41] The next day Mitchell seemed less tense, more combative, and ready for Gullion, who was determined to press his case to a conclusion. The night before, Reid and the defense team had searched the transcript for statements Gullion might return to. After each Mitchell statement, many of which were quite long, the defense left the room for Mitchell to write his thoughts and comments. It was the sort of attention to detail that Mitchell did not care for, and often he did not respond to the points raised by the defense.[42] A number of people around Mitchell, including Reid and Henry "Hap" Arnold, did not think that Billy Mitchell took the court-martial seriously enough.

Certainly Major Allen Gullion was taking the trial seriously, and he was determined to show that Mitchell was either a dangerous self-promoter or a mental case. According to his plan, Gullion had prepared the groundwork and was now ready for his most devastating attack. Gullion returned to a statement that Billy Mitchell made on September 5, 1925, at the opening of the trial, in which he claimed that "the airmen themselves are bluffed and bulldozed so that they dare not tell the truth in the majority of the cases." Gullion asked Mitchell whether he was referring to air officers of the army or the navy, to which Mitchell replied, "I refer to myself principally." Gullion had prepared a trap and Mitchell talked himself into it with ease. Gullion then asked in a voice dripping with sarcasm if Mitchell meant that he rarely told the truth. Billy Mitchell stumbled badly, and Gullion delivered another stinging blow when he repeated the statement that the airmen are "bluffed and bulldozed" and then asked in a louder voice, "Do you consider yourself in the plural habitually?" Reid, seeing the damage being wrought, was on his feet objecting, demanding that Gullion read the entire statement that referred to "bluffed and bulldozed." This was a tactical error because Gullion was ready to oblige by reading the entire paragraph to the court.

Gullion finished reading and then, drawing closer to Mitchell, asked, "Do you mean to imply that you and other officers dare not tell the truth before superior officers and Congressional committees?" Mitchell responded that while he told the truth, other officers, for fear of their careers, were afraid to. "Are those other officers when called before a Congressional committee and asked questions, guilty of perjury or false swearing?" Gullion asked, and Reid, seeing the bloodletting getting worse, was again on his feet, yelling at Gullion that his question was improper in the extreme. Gullion, usually in complete control of his emotions, showed his anger and hollered back at Reid, "It is his own statement!" He then noted that officers are continually on Capitol Hill before committees. Do they, or do they not, tell the truth to Congress? Blanton Winship watched this bitter, loud exchange and saw the reporters taking copious notes. He intervened, telling Gullion that it really was not a proper question, and asked him to move on. Allen Gullion was prepared to do just that, having made his point.

Gullion had opened a severe wound, and he was ready to cut even deeper with another line of questioning. But Reid, who was trying not to be out-lawyered, fell into another trap when he said that the whole paragraph of the previously cited statement declared that if they do not tell Congress what the army wanted them to they were "sent to out-of-the-way places." Yes, replied a very satisfied Gullion, that is exactly what the statement said, and then he asked Mitchell if he had indeed been sent to such a place? Did Mitchell not know that his superior, Major General Mason Patrick, had stated that nobody had been exiled to a far-flung and isolated post? Mitchell weakly responded that he did not know that Patrick had said that. Was Fort Sam Houston such an out-of-the-way post for a soldier? Gullion asked. Because so many officers and troops were stationed there and because there Air Service pilot and ground training were so prominent, why would Mitchell consider it such a place of exile? "I certainly consider it an out-of-the-way place so far as influencing air service development is concerned," Mitchell answered. Gullion then leaned forward toward Billy Mitchell and asked, "How about as far as publicity is concerned?" Reid was horrified at what was happening to Mitchell, but he was powerless to stop it. Gullion had milked this line of questioning as much as he wanted, and he portrayed Billy Mitchell in the worst possible light. As the day came to an end it was clear that the cross-examination of Billy Mitchell had been a success for Gullion and the prosecution.[43]

Guillion did not wait long the next day to go after Mitchell again, trying to show him as a self-promoter. He produced a copy of *Winged Defense* and claimed that Mitchell had plagiarized part of it from a naval officer's lecture. This was a critical moment for Billy Mitchell, and Reid knew it even though Mitchell seemed unaware of the potential damage of such testimony. It reflected upon the integrity of Mitchell and damaged his credibility as a spokesman for airpower. In late November, Gullion had received a memorandum from Major Kyle Rucker of the Operations Section of the General Staff that outlined the glaring similarities between *Winged Defense* and a lecture delivered by Captain T. C. Hart of the navy to the Army War College in 1919 on submarine warfare. At this time the lecture was still classified as confidential. Rucker had been detailed to do as much research as possible to assist the prosecution, and he had documented, page by page, possible plagiarism by Mitchell.[44] Gullion also had in his possession an official objection by Hart to Major General Hanson Ely, the commandant of the War College, sent in March 1925. Hart's detailed objection claimed that Mitchell had copied his ideas verbatim and had then published still-classified material not only in *Winged Defense* but in a paid-for article entitled "How Should We Organize Our National Defense," which appeared in the *Saturday Evening Post*.[45] General Ely had examined the complaint and the lecture, book, and article, and concluded that it was "certainly a most serious offense."[46]

When Gullion asked if Mitchell had written *Winged Defense,* neither Reid nor Mitchell had any idea what was to follow. As Gullion probed, Reid became concerned that Gullion had another motive for asking these questions, and he was on his feet asking what Gullion was doing. At that point Gullion produced the Hart lecture and accused Mitchell of "cribbing" material on submarines from a still-classified document. Gullion thrust the book at Mitchell, asking him to read the marked passages, and Reid exploded, demanding that if Allen Gullion wanted it read he should do it himself. Fine, responded Gullion, who cited a sentence in the book by Mitchell to the effect that everything he said in the book was his own and that he wrote the truth. It was already clear that Gullion wanted either to have a case within a case or to prepare the groundwork for a second court-martial based on the unauthorized and public use of classified materials, a clear violation of regulations and the Articles of War.

Blanton Winship saw the problems with this line of questioning and did not want the original charges and specifications of this court-martial to be sidetracked by another serious offense that could be more properly pursued at a later date by the army. Winship sustained Reid's vociferous objections, but he left the door open for some future legal actions. Gullion was not through, however, and asked Mitchell again if he had written *Winged Defense,* to which he replied in the affirmative. Reid was back on his feet objecting, but not before Gullion asked if any credit was given to any other author or source in footnotes or in a bibliography, to which Mitchell replied no. Winship intervened, telling Gullion again that the court would not deal with this matter any further. Howze, now furious, warned Gullion that the Law Member had ruled on the inadmissability of the book, lecture, and the article. Gullion went on despite Howze's warning, until a very angry Howze, in as strong a voice as possible, ended the whole business.[47] But the damage had been done, and Major Gullion had succeeded in painting a picture of Mitchell as an officer who, for pay, used a classified document for his own glorification with no regard for giving due credit to a fellow officer of another service. The generals of the court were West Point men who believed in Duty, Honor, Country, the motto of the Long Gray Line, and certainly this accusation, even though ruled not admissible, would weigh on their minds. Gullion had won a very important point despite Winship's ruling and the obvious anger of Howze. After constant probing, his questions and comments often dripping with sarcasm, Guillion finished his cross-examination, and Mitchell was excused. Reid then called several other witnesses, including Congressman Fiorella LaGuardia of New York. Gullion asked if LaGuardia had made a statement that Mitchell was not being tried by his peers but by officers prepared to find him guilty, as the General Staff wished. The New Yorker responded that when he made that statement to the press he did not know that General MacArthur was on the jury.

Much to Howze's undisguised disgust, the prosecution stated that they would have more witnesses to rebut some of Mitchell's testimony, including Hugh Drum, Dennis Nolan, and Admiral William A. Moffett. The trial's unexpected length and bitterness were wearing on everyone. Before the Thanksgiving recess, and again on December 4, Mitchell requested permission to visit Boxwood.[48] Permission was granted by the Military District of Washington, and with admonitions from Reid and others to be quiet, Mitchell spent two restful weekends away from Washington. After Thanksgiving, when the trial resumed, Reid questioned in detail the rebuttal witnesses. After some especially tedious cross-examination by Reid, Brigadier General Edward L. King simply blurted out, "damned rot!" Howze knew what was coming and issued an immediate apology, as did King, and Reid did not press for a mistrial.

If Howze thought that the "damned rot" crisis had convinced the court to be judicious in their comments and reactions, he was soon shown to be in error. A few days after General King's outburst, Major General Sidney Graves became embroiled in a clash with Frank Reid. During the cross-examination of navy Captain H. E. Yarnell, former commander of naval air in the Hawaiian Islands, Reid bore in on Yarnell's refusal to include army air in a training exercise in 1924. Gullion objected to Reid's line of questions, arguing that Yarnell's decision was made on strictly military lines. There was a sharp interchange between Reid and Gullion, and Graves turned to a fellow court member and said loudly, "This wrangling is disgraceful and ought to be stopped." Reid exploded and shouted at Graves that he would conduct his cross-examination as he saw fit and did not need any interference. Graves was on his feet, shaking his fist at Reid, saying, "I claim the right to express my opinion to my fellow-member without being subjected to criticism." Howze and Winship were now at a loss as to how to deal with the clash. Reid immediately demanded that Graves be dismissed from further duty with the court and leave the Emory Building. General Howze then cleared the courtroom and spent almost an hour trying to figure out how to deal with the new crisis and avoid a mistrial or the dismissal of Graves from the court. The best Howze could do was to reconvene the court, order the prosecution and defense lawyers to stand, and then chew out everyone. "The Court," Howze growled, "wishes to state that it has viewed with disfavor the constant bickering between counsel for both sides in violation of the court-martial ruling that all remarks should be addressed to the court." Graves, having been instructed by Winship, then said that he had no intention of interfering with the court and felt that he could indeed render a fair verdict in the case. With Howze shaking his head in the affirmative, he indicated that the issue was thus resolved. Graves's comment had come near what seemed to be the end of the proceedings, and so Howze was determined to avoid a mistrial.

Before calling the big three—Drum, Nolan, and Moffett—Moreland and Gullion decided to bring Major Generals Hanson Ely and Robert Allen to the stand for the prosecution. Reid had been making much of accidents in air training, trying to prove what Mitchell had been saying about the outmoded aircraft and dangerous training practices. Bringing Ely to the Emory Building was a wise move because Ely had an excellent reputation among the officers of the army. A 1887 graduate of West Point, Ely had gone to France in 1917 as commander of the 28th Infantry Regiment of the 1st Division. In May 1918, Ely commanded his regiment in the first American combat attack at Cantigny, and his regiment took their objectives, showing that the AEF had striking combat power. Pershing had gambled much on the ability of the regiment to win the fight at Cantigny, because if the attack failed, the possibility of the AEF never coming into existence was great. Ely's reputation was made, and the generals of the jury knew his background. He arrived at the court-martial as commanding general of the U.S. Army War College. He stated that the War College did indeed have lectures on air combat and the Air Service, and graduates left the college with as much information about the air as any other branch of the army. Reid went after Ely on cross-examination but was not able to shake his testimony. At one point, Ely reminded the civilian Reid that the army was made up of many branches, giving the impression that Reid really did not have a firm grasp of army organization. Robert Allen was in 1925 chief of the Infantry Branch, the fount of knowledge for the army at that time. His testimony centered on accidents in training, and he was able to refute claims that the Air Service faced greater dangers than other soldiers in training, handling ammunition, and so on. Reid stated that in 1924 there had been 256 accidents in the Air Service, and he asked Allen to confirm this number. Yes, responded Allen, it was dangerous work to fly in the air. That was why the army gave pilots flight pay, because it was risky. They were compensated for their efforts, and, by the way, as a qualified army aviator Mitchell drew flight pay knowing full well the dangers of flying. Looking at Reid, Allen did say that there was a difference between an infantryman with his feet on the ground, not being fired at in training, and the pilots who flew in the airplanes of the day. Hanson Ely and Robert Allen prepared the way for the testimony of Hugh Drum by showing that Reid really did not have solid information about the organization of the army or exactly how they trained in peacetime.

Before Dennis Nolan and Admiral Moffett testified, the prosecution put on its star rebuttal witness, Hugh Drum, who was primed and ready, having begun preparing for his appearance in November. The prosecution felt that Drum would be one of their most effective witnesses in refuting what Mitchell and the defense had claimed throughout the trial. Drum was highly articulate and had an ability to absorb and retain details and facts. Since his service with Pershing and the AEF in the Great War he had been one of the army's best spokesmen for John J. Pershing's belief in the primacy of the infantry on the battlefield. Along with Major General Dennis Nolan, who would also testify, he was one of Mitchell's most severe critics. Drum had directed the G-3, the Training and Operations branch of the General Staff, to examine Mitchell's statements and prepare rejoinders based on events and on regulations. Colonel Charles S. Lincoln, chief of the branch, threw himself into the task with great results. For example, Mitchell had claimed, "All aviation policies, schemes, and systems are dictated by the non-flying officers of the Army or Navy who know nothing about it." Moreland and Gullion had decided to question Mitchell about this statement when Gullion did his cross-examination, and Lincoln made certain that Drum had chapter and verse from the regulations and from Pershing's guidance when he was chief of staff. The prosecution considered Drum to be an expert witness who could cite definite facts and exact figures with facility.[49] Moreland believed correctly that the prosecution could contrast Drum with Mitchell, who could answer only in vague responses, constantly using such terms as "my opinion" or, worse, "I don't recall."

As the time grew near for Drum to testify, Moreland and Gullion scripted their questions and Drum's responses. In several sessions with the prosecution the questions were refined and Drum's answers made more specific. Redundant questions were eliminated because Moreland had observed the obvious irritation that many of the generals on the court had shown with tedious, repetitive testimony. Moreland penciled in a new question that he knew would draw the attention of every general on the court. He was to ask Drum, "Then in your opinion the criticisms of the defense on the principles, doctrines, and policies system are in reality an attack upon General Pershing?"[50] Every major general sitting on the court, with the exception of Douglas A. MacArthur, owed his combat command and his stars to John J. Pershing. This question—in fact, Drum's entire testimony—would be a defense of Black Jack Pershing, who had near godlike status among the officers of the United States Army. To question the wisdom and the guidance of the General of the Armies was as close to heresy as a professional soldier could get. Drum was the type of witness that Mitchell was not, willing to spend long hours in preparation for his questioning by the prosecution and by the defense. And Hugh Drum would actually be speaking for the retired Black Jack Pershing.

Moreland and Gullion both wanted to show that Mitchell was a shameless self-promoter who was willing to stretch the truth when he wrote or spoke. Mitchell had challenged the propriety of Charles P. Summerall's position as president of the court-martial, and Moreland wanted documentation to show that conditions in the Hawaiian Military Department were not as dire as Mitchell had claimed—and, in fact, that improvements were constantly being made by Summerall and his Air Service commanders after the Mitchell inspection tour in 1923. The staff of the chief of the Air Service combed their files and came up with a number of documents, the most important of which were monthly activity reports for the Air Service for the years 1923 and 1924 which had been endorsed and sent by Summerall to Mason Patrick's office. They showed a pattern of continual improvement in training for the service, with numerous Air Service field exercises. The monthly report for January 1924 contained a telegram from Mitchell to Summerall that stated, "Pursuit and bombardment splendid maneuvers yesterday. Many thanks and aloha to yourself and Mrs. Summerall."[51] A similar telegram had been sent to the Air Service commander in Honolulu. While Billy Mitchell was writing "splendid" to Charles Summerall, he had been preparing a very different document for Mason Patrick. How much of this the court would allow was a question yet to be answered, but the potential damage to an already hemorrhaging defense was great. If they decided to use the unclassified documents, they could very well reveal Billy Mitchell as a back-stabber, the type of officer who did not take loyalty to his superior and to the army as an institution seriously. If Moreland and Gullion felt that they had to use such information, they would, but their case was already a strong one.

Even though the press and Billy Mitchell's supporters had publicly denounced Drum as a relentless critic of Mitchell and an anti-Air Service officer, every point the prosecution wanted to address, except for Mitchell's dealings with Summerall, was covered during Drum's testimony. Dressed in immaculate uniform with ribbons and decorations, Hugh Drum pounded away for two days at Mitchell, refuting much of what Mitchell had claimed, including the widely held belief that Mitchell had command of the Air Service in the AEF. Drum stated that Mitchell never did accomplish what he claimed to during the Meuse-Argonne, and he carefully avoided mention of Billy Mitchell's success of air concentration in the St. Mihiel operation.

While Drum was obviously well prepared to give his testimony and was in command of his facts, two days on the stand began to take their toll. This became clear as Frank Reid began his tedious cross-examination. Drum had carefully cultivated his image as a highly professional soldier, probably in part because he did not wear the West Point ring, and became irritated with Reid's clumsy civilian handling of military matters. When asked about the current organization of the Air Service within the War Department and the General Staff, Drum became angry and told Reid that the present organization was created by General Pershing himself. Reid went on and Drum became more hostile, asking Reid just exactly what were Reid's "war experiences." Do you recall the effects of anti-aircraft fire on missions? Reid demanded of Drum. Drum replied that he did not recall exact missions, and he then asked Reid if he personally recalled any incidents. By this time the spectators were on their feet, and the military police guards were eyeing the crowd with some nervousness. Mitchell jumped to his feet and shouted at Drum, "I was as much in my line during the war as you were in yours, and I wasn't a chief clerk of the General Staff either." Now Drum was out of his seat, and it appeared that Mitchell and Drum would roll up their sleeves and fight each other right there in the courtroom. Moreland and Gullion restrained Drum, and Reid grabbed Mitchell quickly. The hatred between the two officers infected the whole crowd, and Howze, concerned about a general brawl breaking out in the Emory Building, cleared the court. After about half an hour the trial resumed with Howze, in his best parade round voice, warning the participants and the spectators that there would not be a repeat of the outbreak of emotion.[52]

It was obvious that Hugh Drum was the star rebuttal witness, and Moreland and Gullion had every reason to be pleased with the effect that

"Drummie," as he was known to his fellow officers, had on Howze and the court. Dennis Nolan and Admiral William A. Moffet were used simply to reinforce Drum's testimony. There was no way Nolan or Moffett could hide their hostility toward Billy Mitchell. When Moffett had read of Mitchell's San Antonio statement, he had exclaimed, "That son of a bitch is riding over the Navy's dead to further his own interest!"[53] Appearing before the Morrow Board and the *Shenandoah* court of inquiry, Moffett showed his feelings, and his appearance at the Mitchell court-marshal was basically a restatement of his opinions. What Moffett did was to show Mitchell as being woefully short of knowledge about naval aviation. Once Nolan and Moffett were excused from the court, Mason Patrick took the stand and said little to help or hurt Mitchell. For four years Patrick had tried to protect Mitchell from himself with overseas inspection tours and trips throughout the United States, but now he could do very little, and his sadness showed.

Reid was not pleased with the prospect of ending the trial with no closing statement of his own, and he was even less happy with Mitchell's request to make a personal statement to the court. Billy Mitchell certainly had the right to address the court, but Colonel Winship told Mitchell that he had the right to remain silent, to be sworn again as a witness, or to make an unsworn statement that would not be subject to cross-examination. Mitchell replied that he wanted to make a brief comment. Winship then asked Mitchell if he wanted to confer with Reid before giving his statement, and Mitchell, with no hesitation, told Winship that Reid already knew what he wanted to do. Gullion offered no objection, and so Mitchell began: "My trial before this court-martial is the culmination of the efforts of the General Staff of the Army and the General Board of the Navy to deprecate the value of air power and keep it in an auxiliary position which absolutely compromises our whole system of national defense."[54] Reid had a worried look on his face as he watched Allen Gullion taking notes and smiling. Frank Reid had no prior knowledge of exactly what Billy Mitchell might or might not say to the court, and it was now clear to the congressman from Illinois that his old friend was determined to alienate the court and seal his fate.

Mitchell claimed that there was a governmental conspiracy, begun in 1919, which totally ignored the *Ostfriesland* sinking and other tests on naval ships. "The truth of every statement," Mitchell claimed, "which I have made had been proved by good and sufficient evidence before this court, not by men who gain their knowledge of aviation by staying on the ground and having their statements prepared by numerous staff to bolster up their predetermined ideas, but by actual flyers who have gained their knowledge first hand in war and in peace." Mitchell then asked the court to look at Secretary Weeks's letter to Coolidge asking that Mitchell not be reappointed to his position as assistant chief of the Air Service. Weeks did this, Mitchell claimed, because Mitchell had testified before Congress. Looking directly at Howze, he concluded his statement, saying in a loud voice that could be heard by all, "The Court has refrained from ruling whether the truth in this case constitutes an absolute defense or not. To proceed further with the case would serve no useful purpose. I have therefore directed my counsel to entirely close our part of the proceedings without argument." Gullion, who looked like the cat who had finally cornered a canary, jumped to his feet and asked Reid if that was a sworn or unsworn statement. Winship understood that Gullion wanted Reid to say that what Mitchell had just said was sworn testimony, and he knew that Gullion would love nothing more than another chance to confront Mitchell. As quickly as he could, cutting Reid off, Winship ruled that Mitchell had given an unsworn statement, thereby avoiding the possibility of lengthy and further damaging cross-examination.[55]

No sooner had Winship ruled on Mitchell's statement than Gullion rose to make his closing remarks. It was so fast that Colonel Moreland had to intervene and request the court's permission for the prosecution to make their final arguments. Howze quickly agreed, and Gullion began with a scathing refutation of everything that Mitchell and the defense had presented. At the end of his lengthy summation, Gullion went after Mitchell as an officer, asking:

> Is such a man a sage guide? Is he a constructive person or a loose talking megalomaniac, cheered by the adulation of his juniors, who see promotion under his banner, and intoxicated by the ephemeral applause of the people whose fancy he has for the moment caught? Is this man a Moses, fitted to lead the people out of the wilderness which is his own creation, only? Is he of the George Washington type as Counsel would have you believe? Is he not rather of the all too familiar and demagogue type—and except for a decided difference in poise and mental powers in Burr's favor, like Aaron Burr? He is a good flyer, a fair rider, a good shot, flamboyant, self-advertising, wildly imaginative, destructive, never constructive except in wild non-feasible schemes and never overly careful as to the ethics of his methods.[56]

It was Gullion at his best in the courtroom, and his summation was devastating. Howze and the generals of the court had listened to every word, obviously balancing this presentation against Mitchell's short, ineffective, and petulant closing statement. Visibly relieved, Howze rose and informed the courtroom that the generals would now retire to decide the guilt or innocence of Billy Mitchell, as if that fate was ever really in doubt.

At 7:25 P.M., December 18, 1925, a courier handed a memorandum to Brigadier General Samuel Rockenbach, commander of the Washington Military District, from Colonel Sherman Moreland, which said, "You are notified that the General Court-Martial appointed by Special Orders No. 248, War Department, October 20, 1925, has found Colonel William Mitchell, Air Service, guilty of the Charge and all Specifications preferred against him and has sentenced the accused to be suspended from rank, command and duty with forfeiture of all pay and allowances for five (5) years."[57] It was not an unexpected verdict. When Billy Mitchell rose to face General Howze to hear the verdict and sentence, the chief of staff of the Army, John L. Hines, Pershing's handpicked successor, was playing golf in St. Augustine, Florida. Many of the reporters had already written their headlines and lead paragraph. The humorist Will Rogers, who had visited Mitchell in the courtroom, was already in New York lamenting the verdict. Someone remarked that the only person he saw who looked surprised and downcast was Major General Mason Patrick.

NOTES

1. War Department, Information Release, 20 October 1925 (with marginal notes), in the William Mitchell Papers, Library of Congress, Washington, DC, Carton 38. (Hereafter cited as the Mitchell Papers.)

2. War Department, Special Order No. 248, 20 October 1925, Mitchell Papers, Carton 52.

3. Moreland to Chief of Staff's Office, ca. October 1925, in National Archives, Archives II, College Park, MD, Records Group 153, Records of the Judge Advocate General of the Army, General Courts Martials, William Mitchell, Case No. 168771, Carton 9214-20. (Hereafter cited as RG 153.)

4. *New York Times*, 29 October 1925, 2.

5. Ibid., 29 October 1925, 1.

6. Statement Prepared by Mitchell on the Challenge to Summerall, Final Draft, completed probably 28 October, Mitchell Papers, Carton 38.

7. *New York Times*, 29 October 1925, 1–2.

8. Ibid., 30 October 1925, 1–2.

9. Ibid., 31 October 1925, 1, 10.

10. Ibid., 13 October 1925, 1. Also see James J. Cooke, *Pershing and His Generals: Command and Staff in the AEF* (Westport, CT: Praeger Publishing 1997), 49–50.

11. Notes and Marginalia to Mitchell's Statement, by Mitchell, late October 1925, Mitchell Papers, Carton 37.

12. Burke Davis, *The Billy Mitchell Affair* (New York: Random House, 1967), 237. Davis indicates that it was Bissell and Webb who located witnesses. It was Mitchell who did the original list for the points he and Reid wanted to defend.

13. *New York Times*, 2 November 1925, 8.

14. Document, "Opening Statement: Points Suggested from the Air Service Point of View," ca. early November 1925, RG 153, Carton 9214-15.

15. The coupons, letters, and petitions are contained in Records Group 407, Records of the Adjutant General, Files, National Archives, Washington, DC, Carton 1681.

16. "Aviation and Colonel Mitchell," *The Independent* 115 (September 1925), 314.

17. Hersey to Hines, 5 October 1925, in the John L. Hines Papers, Library of Congress, Washington, DC.

18. Hines to Hersey, 12 October 1925, Hines Papers.

19. *New York Times*, 3 November 1925, 1.

20. Ibid., 4 November 1925, 4.

21. White to Moreland, 21 November 1925, RG 153, Carton 9214-6.

22. Memorandum by Gruelick, 18 November 1925, RG 153, Carton 9214-6.

23. Article, ca. November 1925, contained in the Hardie Collection, Golda Meir Memorial Library Archives, University of Wisconsin, Milwaukee, Scrapbook 5.

24. Memorandum (Secret), Aerial Photographs, 22 November 1924, Hardie Collection, Scrapbook 5.

25. *New York Times*, 12 November 1925, 1–2.

26. Davis, *Mitchell Affair*, 271.

27. *New York Times*, 14 November 1925, 1–2.

28. "Notes of Sources of Information and Cross Examination," prepared by Gullion, RG 153, Carton 1925-12.

29. RG 153, Carton 9214-4.

30. Ibid., Official Transcript, pp. 2190–2211.

31. *New York Times*, 16 November 1925, 5.

32. Ibid., 17 November 1925, 10.

33. Samuel Taylor Moore, "Airplanes Land in the Field of Politics," *The Independent* (19 December 1925), 700–702.

34. *New York Times*, 19 November 1925, 8.

35. *The Congressional Digest* (April 1925), 241, 247. Mitchell also contributed several position articles to this edition devoted to the airpower debate.

36. *New York Times*, 15 October 1925, 1.

37. Transcript of Rickenbacker's Testimony, 19 November 1925, in the Edward Rickenbacker Papers, Library of Congress, Washington, DC, Carton 20.

38. Position Paper from G-3, General Staff to Gullion, undated, ca. early November 1925, RG 153, Carton 9214-4.

39. *New York Times*, 24 November 1925, 1, 2.

40. Letters to the Editor, *The Independent* (4 April 1925), 391.

41. *New York Times*, 24 November 1925, 2.

42. Memorandum and Transcript of Testimony, Prepared for Mitchell, Mitchell Papers, Carton 37.

43. RG 153, Official Transcript of Testimony, prepared for Mitchell's response, Mitchell Papers, Carton 9214-3.

44. Rucker to Gullion, 23 November 1925, RG 153, Carton 9214-15.

45. Hart to Ely, 20 March 1925, RG 153, Carton 9214-15.

46. Endorsement by Ely, 20 March 1925, RG 153, Carton 9214-15.

47. RG 153, Official Transcript, pp. 1568–1573, Carton 9214. *New York Times*, 25 November 1925, 1.

48. Mitchell to Rockenbach, 25 November 1925, and Mitchell to Rockenbach, 4 December 1925, Mitchell Papers, Carton 11.

49. Lincoln to Drum, 21 November 1925, RG 153, Carton 9214-12.

50. "Drum, Trial Handling of, Questions for . . . ," RG 153, Carton 9214-15. The pencil notations, additions, and deletions by Moreland are numerous and show a changing prosecution strategy. *New York Times*, 25 November 1925, 1.

51. Activity Report for Month of December 1923, Air Service, dated 3 January 1924, RG 153, Carton 9214-12. *New York Times*, 16 December 1925, 16.

52. *New York Times*, 17 December 1925, 8.

53. William F. Trimble, *Admiral William A. Moffett: Architect of Naval Aviation* (Washington: Smithsonian Press, 1994), 160.

54. RG 153, Official Transcript, pp. 3691–3694, RG 153, Carton 9214-5.

55. Ibid., pp. 3694–3995.

56. Ibid., pp. 3742.

57. Sherman to Rockenbach, 18 December 1925, Mitchell Papers, Carton 11.

ELEVEN

RESIGNATION AND CRUSADE

T HE GUILTY VERDICT IN THE COURT-MARTIAL OF BILLY
Mitchell had been expected by everyone, with the possible exception
of his most rabid supporters. The fallout from the trial was loud and bitter,
and nowhere was that more evident than in Congress, where the upcoming
congressional electoral campaign was a source of much posturing and
position-taking over Mitchell, the Coolidge administration's views on air,
and the condition of the U.S. armed forces (not that there was much interest
by either party in substantial increases in the military budgets). Some
Democrats in Congress warned that they would take steps to overturn the
Mitchell verdict, propose air legislation of their own, and make Mitchell
head of a unified air service. The problem for those politicians was that
they were in the minority in Congress, and President Coolidge's Morrow
Board had issued a final report that was moderate, proposing steps that
were workable. One of the consistently cooler heads amid the welter of
words over the Mitchell trial was Mason Patrick. In late November 1925
Patrick sent a report to Secretary of War Dwight F. Davis that called for the
establishment of an Air Service having about the same relationship to the
War Department that the U.S. Marine Corps had to the Navy Department.
This relationship would exist until a proper "Ministry of Defense" could be
established. This Air Service Corps would have a separate budget that
would recognize the special needs of the air.[1] Omitted in Patrick's report
was any mention of naval aviation. Patrick's reasoned approach and his
estimation that much had been accomplished in the Army Air Service in a
year won him editorial praise from the *New York Times* and other major
newspapers.[2] This was in sharp contrast to the shrill proceedings in the
Emory Building in Washington, where Mitchell's trial had reached about
midpoint.

221

On December 2 President Coolidge released the findings and recommendations of the Morrow Board, which flatly rejected many of Mitchell's more extreme claims about airpower. The board recognized reality: the Air Service was really a very young component of the national defense and would not, until technology improved, be the decisive factor in war. Even so, the Air Service had the potential to be a powerful weapon, as Mitchell had shown in the *Ostfriesland* and *Alabama* bombing demonstrations in 1921. A separate cabinet-level position for defense, with air having an equal voice with army and navy, was not yet justified, however. The creation of secretaries of aviation within the Department of the Army, Navy, and Commerce certainly was something that needed to be done. Seated on the Morrow Board were former Assistant Chief of Staff James Guthrie Harbord, now an RCA executive, and Senator Hiram Bingham of Connecticut, himself a Great War aviator. They knew the needs and they really understood the art of the possible for development of air in 1925.

Older flying officers in the Army Air Service saw the potential in the Morrow report. Colonel Benjamin Foulois, who would become chief of the Army Air Corps in 1931, was one of those who saw the Morrow Board as a step in the right direction. He wrote to a friend after the publication of the board's findings, "In my opinion we will never develop aviation by following the line of attack, which has been so clearly evident, not only during Mitchell's present trial but during the past several years of investigation. I am fully of the opinion that the recent report of the President's board should form a foundation for this winter's legislation, and if we can not evolve some sort of program for advancement of aviation, based on that report, I fear that we will have a very hard time in making future progress."[3] From Dayton, Ohio, the venerated air pioneer Orville Wright issued a statement in full support of the Morrow Board, saying, "It is the best summary of the situation I have seen. Its recommendations are sound."[4] Mason Patrick was relieved that as far as he was concerned, the wrangling was over and small first steps toward independent airpower could now be taken.

It was an irritation for Mitchell to read in the *New York Times* an editorial that stated, "All told, it [the Morrow Board] has furnished Congress and the country with material for clear thought and appropriate action such as we have not had before."[5] But Billy Mitchell had not had the time or inclination to comment on the board while his trial was in progress. Once it was over that would be a different matter. Mitchell was bound by the rules of the court-martial to remain in Washington and to restrict his activities until an army board could review the findings and then send their recommendations to President Coolidge. Over the Christmas season William and

Betty would be in Washington for a while, and he had to apply to General Rockenbach for permission to visit Boxwood during the holidays. This was granted, provided that Mitchell did not take the opportunity to speak to the press. On January 14, 1926, Mitchell made a strangely worded request for permission to see his in-laws leave New York for a vacation in Europe: "I request that I be given permission to visit the city of New York . . . for the purpose of seeing my father and mother off on a trip to Europe."[6] Rockenbach and his staff must have puzzled over the wording—Mitchell's father and mother had been dead for many years. Mitchell was granted only two days in New York, not the five he had requested, with strict admonitions as to what he could and could not do.[7] These restrictions grated on Mitchell, who had to remain on the sidelines while Congress began debating air policy.

Of equal importance was the fact that if he accepted the sentence of the court, a five-year suspension, he was still subject to military regulations that prohibited private employment, lecturing for a fee, and even writing for profit. Several weeks went by as the review board was assembled to look at the 4,000-page transcript of the trial. Frank Reid did not understand the functions of a board, especially one for a general court-martial, and continually complained to the press that the army was making Mitchell suffer in limbo. But Mitchell was not a man to suffer in silence, and, in a bizarre manner, he thrust himself into the newspapers again. On January 19, Mitchell and his wife appeared at a meeting of the House Military Committee, where hearings were underway to see if the creation of a Department of Defense was in the national interest. In civilian clothes, Mitchell asked to testify, and this set off a debate about the propriety of an officer under sentence of court-martial testifying before Congress. Mitchell shouted that he was there voluntarily, and that "My status is no different today than it has been."[8] The committee was dumbfounded, and even Mitchell supporters were taken aback by this. The entire committee, in a bipartisan agreement, decided that it would have to wait for guidance from the War Department before allowing Mitchell to say anything at all. On the same day, the review board upheld the decision of the court and passed the matter to President Coolidge for review.

Billy Mitchell was now the subject of yet another wrangle within the War Department. Frank Reid had decided to make a political case out of the department's decision about Mitchell's right to speak before the House Committee, and he stated that if the court-martial sentence were upheld by Coolidge it would in effect "gag" Mitchell. In a surprise move, Major General J. A. Hull, the judge advocate general of the army, ruled that Mitchell could testify, and, if the sentence were upheld, Mitchell could seek employment in business if he so wished, with some restrictions. On the same day, January 22, 1926, the House cut $9 million for new aircraft construction from the defense bill.

Major General Hull began a careful review of the trial, intending to make a recommendation to President Coolidge. Realizing that the court-martial of Billy Mitchell had captured headlines from coast to coast, Hull was determined to be as fair as possible, presenting the court's verdict as just but tempered with mercy. He believed that the guilty decision was the correct one based on the facts of the case, regardless of Mitchell's defense that his statements, which had brought on the trial, were correct. Hull therefore advised Coolidge as a matter of military law that extenuating or mitigating circumstances were not an issue as far as the president was concerned. "Had there been anything wrong," Hull told the president, "it would have been preferable to bring it to light so correction could be had, rather than to have it suppressed or glossed over."[9] In reading the lengthy four-thousand-page transcript, Hull noted that Mitchell, under cross-examination by Allen Gullion, constantly stated that statements made by the accused were opinions only. As far as the judge advocate general was concerned the fact that Mitchell said them was a violation of Article 96 of the Articles of War. Coolidge, however, urged Hull to lighten the sentence imposed by the court to allow Mitchell to receive some pay and allowances during the five-year suspension.[10]

Coolidge studied Hull's recommendations and decided that he would adopt the well-considered arguments as his response to the court's guilty verdict. The president's statements reviewed the *Shenandoah* crash and cited the desire of the people to know what had caused such a tragedy. In light of this Mitchell had made his statement, accusing the government of "incompetent, criminally negligent, and almost treasonable" activity. Coolidge agreed wholeheartedly with the court and the military review board headed by General Hull that Mitchell's statements were "made without basis in fact." In the president's view, Mitchell had violated the relationship between subordinate and superior that was vital to the maintenance of good military discipline. To soften the financial blow for Mitchell, Coolidge allowed him to draw half pay and allowances, and he did this upon the recommendations of the judge advocate general and the Secretary of War.[11] There were very vocal critics of the court-martial and Coolidge wanted to defuse any further anger over the verdict by showing that the highest law officer in the army, the secretary of war, and the chief executive could be fair and aboveboard. It did not deter Frank Reid from denouncing Coolidge's decision as "un-American."[12]

But Reid was not through with the War Department, the General Staff, or Calvin Coolidge. Conspiracy! Reid charged that Coolidge had acted upon the case within the space of a few hours (which was not accurate by over a week), and he asked, "Why this unseemly haste?" The whole case from investigation to verdict, coupled with the alleged speed by Coolidge, "shows that an invisible mind and hand have labored without ceasing for this day's verdict. *They* [my emphasis] have finally got Colonel Mitchell."[13] Of course, Reid was the lead defense counsel and had, for many years, believed in Billy Mitchell, but it was Mitchell who brought himself to this result. Mitchell then decided that his best course was to resign from the army as of February 1.[14]

Mitchell had twenty-seven years in the army, and he needed three years to draw his pension upon retirement. He would not be able to return to his rank as a colonel until 1930 and would not have his thirty years until 1933. He would, at half-pay and all allowances, still be subject to military regulations and restrictions, but the recent ruling by General Hull left the door open for Mitchell to speak and to enter into business. Frankly, the army was generous with Mitchell, but Mitchell's ego would not allow him to accept the verdict of the court-martial.

On February 1, Mitchell called a press conference to make a statement explaining his resignation from the army. He started out by citing his service from the 1st Wisconsin Infantry through the Great War. Mitchell did claim that he had "command of our air forces in the World War and the privilege of directing the greatest concentration of allied air power." While he conveniently forgot Mason Patrick as chief of the AEF's Air Service, he was on target as far as the St. Mihiel operation. Accusing the military bureaucracies of blundering, failing to come up with a coherent air policy, and destroying morale, Billy Mitchell went on to say, "They have coerced, bulldozed, and attempted to ruin patriotic officers who have disagreed with their views and who sought to better our national defense which the officers and men of the armed services know is necessary." He ended his statement with, "From now on I feel that I can better serve my country and the flag I love by bringing a realization of the true condition of our national defense straight to the people, than by being muzzled in the Army. I shall always be on hand in case of war or emergency, wherever I am needed."[15] There was no question as to what Mitchell might say as a civilian, but what venue it would take—lecturer, writer, senator—was another matter indeed.

There was much dithering in Congress over the sentence handed down in the Mitchell case. Several wise solons suggested that Generals Howze, Nolan, and Drum should receive the same five-year suspension from rank and full pay. It seems that the fact that those generals were not on trial for any violation of the Articles of War was lost on them. Congressman Charles

F. Curry, the author of the Curry Bill, called a press conference to announce that there was a move in the House to place Mitchell on the retired list as a major general, with all the retired pay to which that rank was entitled. According to the congressman, Mitchell had been "forced out of the army as a punishment for telling the truth."[16] The *New York Times*, which had carried all of Mitchell's travails in 1925 with fairness and balance, editorialized that the president's actions were indeed just, and if Mitchell really had the good of the service at heart he would set an example for younger officers by accepting his punishment. The paper blasted Reid's conspiracy theory as an absurdity.[17]

In his typical fashion Mitchell rushed into a lecture tour of the United States, signing an agreement with the Pond lecture circuit bureau in New York. The first few stops were deceptive; in Altoona, Pennsylvania, 25,000 people turned out to greet Mitchell and the lecture hall was full.[18] February and March, however, were not months to be traveling or expecting people to venture from their warm houses, and often crowds were very small and the lecturer's fee equally reduced. The year 1925 had been filled with stories about the air controversy, but that did not mean that the public would retain their interest for very long or that the court-martial, now over, would be enough to entice the spending of a few dollars to hear Billy Mitchell. The American Legion post in Chicago, which was sponsoring a Mitchell appearance, became alarmed at the small number of ticket sales. The enterprising Legionaries announced that there would be a prize of tickets for the fall University of Illinois versus University of Chicago football game, where the famed Red Grange was to play.[19] Of course there were a vast number of invitations for Mitchell to speak at civic organizations for free, just for a plate of cold chicken, hard peas, and rubbery mashed potatoes. Mitchell, now without an army salary and with mounting debts, had to concentrate on his lecture tour.

The sales of *Winged Defense* during the hectic year of 1925 should have been a warning to Mitchell. Before his resignation George Putnam, owner of the company that published the book, sent Mitchell a check and a note saying, "Here's the account and a check. Sorry the latter is not larger. I can't remember a non-fiction book which in one way or another stirred up so much attention and at the same time sold so disappointingly." The check was for $310 (an advance of $1,000 had already been paid) for the sale of close to 4,500 books, and Mitchell had bought forty-nine of those.[20] The public, whose taste for the sensational could be fickle and who also were tightfisted, as the tour was showing, now found Billy Mitchell less interesting—he was becoming yesterday's news.

There were, nonetheless, individuals who rushed forward to take advantage of Mitchell's fame and notoriety with every possible scheme. A Springfield, Massachusetts, land speculator bought an island of 712 acres near a naval fleet oil station in Hampton Roads, Virginia. Mitchell was told the island was great for an airfield, and that he had tried, unsuccessfully, to sell it at $500 an acre to either the army or navy. If Mitchell would use his friends in Washington to good advantage, the land speculator would give Mitchell 2 percent of the final price.[21] Billy Mitchell, to his credit, never responded to the person even though the need to make money was very much on his mind.

While Mitchell was on his lecture circuit in the snows of New England and the Midwest, pressures from the court-martial could still be felt in the War Department and the office of the chief of the Air Service. Mason Patrick would be charged with preparing the draft legislation for an Army Air Corps, and he stated that he saw the corps as analogous to the relationship between the navy and the marines, very close to a separate status. Secretary of War Davis was not in favor of this approach, despite the fact that Patrick had stated that he favored a unified service at some future point.[22] On January 28, a bill authored by Patrick to create an Army Air Corps was offered on the floor of Congress.[23] While Congress debated the fate of the bill, other plans that would have followed Mitchell's unified air plan were discussed, and there was intense lobbying by civilian groups that favored the Mitchell plan. The U.S. Air Force Association, which claimed to be a "Bureau of Publicity and Information," located in Washington, was one of the main voices for unification. Unfortunately the directors learned too much from Mitchell. The national chairman (and chief backer) was Eddie Rickenbacker, who was almost fanatical in his support for Billy Mitchell. The advisory board consisted of colonels Harold Hartney, Joseph F. Randall, W. G. Shauffler, and Major Harry R. Horton, all of whom saw service under Mitchell in the Great War and were devoted to him. In attacking what would become the Army Air Corps Act, the association stated, "When a Congressional Committee calls for information on matters pertaining to the Air Services of the Army and Navy, there appears on Capitol Hill an assorted aggregation of swivel-chair officers who know as much about aviation and the problems involved as a Choctaw Indian does about Sanskrit." The association went on to charge that the army and navy took special steps to see that no qualified officers testified before a committee.[24]

While this characterization was certainly unfair to Patrick and Admiral William A. Moffett, it did point to the tensions that existed. It also encouraged younger officers to follow in Mitchell's footsteps, and that was unwise. In late February Mason Patrick kicked two promising officers out of his office and reassigned one of them outside Washington. Majors Henry "Hap" Arnold and Herbert A. Dargue, the officer who had taught Patrick to fly at Bolling Field, had printed up a circular supporting Patrick and advocating the Air Corps plan. They then went to Capitol Hill and passed their circulars to members of Congress, and when Patrick found out about it he was furious. The Air Service had had enough of Mitchellesque lonewolf tactics, regardless of the cause. Patrick stated, "Both will be reprimanded and one, Arnold, no longer will be wanted in my office. He will be sent to another station."[25] Immediately Mitchell stated to the press that the disciplining of these officers was another example of muzzling opposition by the War Department, and probably this was his first step in going after Patrick himself. With Patrick's bill being debated and fine-tuned in Congress, now was not the time to have anything or anyone derail the process. Patrick believed in a unified service when the time was right, and he regarded this act, if Congress did pass it and fund it, as a major step in the right direction.[26]

Mitchell made a brief appearance before the House Committee on Military Affairs on February 5, just four days after he resigned. He went to argue for the unified air service and against what Patrick had proposed, and his testimony was sometimes on the verge of being a revelation from a soothsayer. Mitchell warned that wars of the future would be fought in the air, with armies on the ground never seeing or closing with the enemy. He foresaw "aerial torpedoes," dropped from aircraft, traveling to a target at hundreds of miles an hour, carrying a huge payload of munitions. Calling this "the military warfare of the future," Mitchell went on to call for unified services, training, and procurement for the air. Again he lashed out at the battleship navy, arguing that the submarine was the sea ship of the future. Somehow Mitchell had lost track of the development of the aircraft carrier program when he said that the carriers *Langley* and *Lexington* were obsolete and that dirigibles should be the airplane carriers of the future—huge airships. In very typical Mitchell style, when asked by Congressman Louis A. Frothingham of Massachusetts what type of organization was required for the unified air service, Mitchell responded, "You don't have to have a complicated organization. Make it as simple as you can. Most of the bureau people are writing novels to each other which nobody reads but themselves—I mean in these departments."[27] This was vintage Mitchell: speaking in vague generalities, making wild predictions that would prove in the future to be correct.

The army had Mitchell pretty much where they hoped he would be—in the snow and cold on his lecture tour and not roaming the halls of

Congress. President Coolidge had enough of Mitchell as well, and he saw the majority of support in Congress for Mitchell coming from the minority party. Mitchell had been invited to address the Union League Club of Detroit on Washington's birthday. The Detroit newspapers reported that President Coolidge's private secretary had called the club president and talked them into canceling the dinner speech. The U.S. Air Force Association exploded, issuing a press release:

> The action of the White House in interfering with the address to have been given before the Union League Club of Detroit on Washington's birthday by Colonel Billy Mitchell is little short of amazing and is another example of autocracy running wild. The fight being made by Billy Mitchell for the betterment of the National Defense is not a partisan political proposition in any sense of the word. . . . If the President and his advisers desire to make this a political fight, we accept the challenge.[28]

The White House indignantly denied the pressure on the club. It had always been partially a political fight, with Mitchell going back to his constant references to running for the Senate from Wisconsin in 1921. Senator James Wadsworth and others recalled it quite well. Betty harbored political ambitions for Billy Mitchell as well.

When he opened his tour at Carnegie Hall in New York, Mitchell stated that he was opening a campaign to pressure Congress into considering and then creating a unified air arm.[29] But many were unimpressed. The *New York Times* editorial page criticized Mitchell for "insisting on receiving very considerable guarantees of payment before he will deliver a lecture about the present condition of aviation as he sees it. This somewhat detracts from the credibility of the assumption that his only thought has been the interests of his country. . . . For him to go only where he is paid, and paid well, puts his campaign on a wholly different basis from the one it ought to have, and the impression created is a regrettable one even from his own point of view."[30] In Milwaukee, in March, he claimed that he had no intention of running for political office. He was not being honest with the people of his hometown, as politics and political parties were very much on his mind as he crossed the country.

In April Mitchell wrote to his father-in-law, Sidney Miller, still calling him "My Dear Father," about his plans for the future. Miller must have been wondering how his son-in-law was planning to support his wife, and Mitchell told him that he really wanted to return to Boxwood and "handle horses, do a little more writing and lecturing and take trips from time to time." Then Mitchell added,

> Betty, I think, would like to be in the Senate and I am not particularly
> adverse to it, although I would like to have it come three or four years
> later. Of course, if one is going to strike, the time to strike is when the iron
> is hot. The more I see of conditions around the country, the less I think of
> the way the Republican party is handling matters in Washington. . . .
> These things naturally incline me toward the Democratic party. On the
> other hand, if the Democratic party again takes up the League of Nations
> or the World Court, I certainly cannot subscribe to it.

His letter made clear that he was looking at a possible run for the Senate or
the House of Representatives from Wisconsin. He added, "Considerable
financial outlay is required in either case, but this could be raised without
much trouble."[31] Given the fact that Miller had covered Mitchell's considerable trial expenses and that Elizabeth was spending her own funds on
Boxwood, this must have made Sidney Miller wince.

If anyone could know what Mitchell was thinking, it was his adoring
sister Harriet. At the same time he wrote to Miller, he wrote to Harriet, who
was still in residence at Meadowmere. With her he was much more open,
writing, "It would be comparatively easy for me to get the Democratic
nomination in the primaries. About $25,000 would be required for the cam-
paign. The more I see of things around the country, the more I think it
would be better for me to be a Democrat, anyway. You might find out about
this phase of the thing as soon as you can and let me know."[32] Mitchell also
hinted to other friends, such as George Putnam and Eddie Rickenbacker.
While he was telling the good people of Milwaukee that he was not interested in running for office, Mitchell told F. B. Patterson, president of the
National Cash Register Company in Dayton, Ohio, "I am not sure that I can
get away for Africa [he and Elizabeth planned a hunting vacation there] in
May because conditions may make it necessary for me to stay in
Wisconsin, as they want me to run for the Senate; but about that I have not
yet decided."[33]

What sort of candidate would Mitchell have made? Would he have
been a one-issue politician? By 1926 Mitchell had a number of ideas in
addition to his main theme of a unified air service. He believed that the
Republicans were controlled by big business and that power had gravitated
into the hands of the executive branch headed by the president and supported by the bureaucracies. He continually argued for lessening the executive
branch's power and placing authority in the legislative branch. Ironically, if
Mitchell had ever gotten his unified air service and a cabinet-level secretary of defense he would have added another Washington bureaucracy.
Mitchell was totally opposed to Prohibition, calling it "an absolute joke."

He had been a drinker all of his adult life, and had problems with it in 1921. His remedy for Prohibition was to allow light wines and beer for the public. He was opposed to discriminatory freight rates that affected manufacturing in the Midwest, and he believed that the development of waterways in the Midwest should be a national priority. In foreign policy he opposed the League of Nations and the World Court, preferring to maintain the United States, Britain, France, and Italy axis of the Great War. Like a vast number of Americans, Mitchell saw no danger in Italian fascism, and he praised Benito Mussolini for his emphasis on Italian airpower. Soviet communism was an anathema for Mitchell, and he urged that the United States have no dealings with Russia at all. These themes continually surfaced when he campaigned for Democratic presidential candidate Al Smith in 1928, and later when he toured the nation supporting the campaign of Franklin D. Roosevelt. Had Mitchell run for public office from Wisconsin he would have had a consistent set of ideas, but air matters might have overshadowed all else. Billy Mitchell was a much better writer than a public speaker, and that would have been a problem for him.

With all of the speculation about Mitchell's political future and the interest in a Mitchell run for the Senate, what happened to Mitchell? Why did he not take the plunge into electoral politics in 1926? There were those pundits who saw Billy Mitchell's problem in fairly clear terms. *The Outlook* put the matter in focus when it editorialized, "It is true that, in a certain way and to a certain extent, Mitchell has been punished. His reputation has suffered by his conviction, and not so much by his conviction, even, as by publication of the facts which led to his conviction. . . . He has suffered perhaps most of all, by his own act of submitting his resignation before his sentence [had] fairly began, by his failure to be in adversity the good soldier and take his punishment like a soldier."[34] The problem rested within Billy Mitchell himself. He believed that he should be appointed as chief of the Air Service, first in the AEF and then in Washington. To many around him it seemed obvious that Mitchell thought he should be the first secretary of air or secretary of defense.

His exalted and often exaggerated view of his own importance was such that a run for the House of Representatives was never in the cards, and facing a determined opponent on the stump for Senate was unappealing. When he campaigned for Al Smith and for Franklin D. Roosevelt rumors surfaced immediately that if those Democratic presidential candidates won Mitchell would be appointed to a post in the cabinet. It was safer for the ego, for his view of himself, that Mitchell remain the martyr—for airpower, for national defense—the sportsman, the gentleman Virginia horseman, writer and lecturer, and bugaboo in the committee rooms of Capitol Hill, reminding lawmakers of the rightness of his own cause. There could never really be a Mitchell candidacy for the Senate.

Billy Mitchell was a tireless self-promoter in 1926. He sent copies of *Winged Defense* to Cecil B. DeMille in Hollywood and to Will Hays, president of the Motion Picture Producers and Distributors of America in New York.[35] He remained in constant communication with George Putnam, who had initially advised Mitchell against the lecture tour. Mitchell asked Putnam if he should revise *Winged Defense*, adding new post-trial material, or whether he should write a new book on aviation, which he assumed Putnam would publish.[36] *Collier's* magazine paid Mitchell $1,000 for an article, "When the Air Raiders Come," which appeared on May 1, 1926. But some of Mitchell's friends, like George Putnam, were becoming weary of Mitchell's constant projects from which he hoped to make a good deal of money. Mitchell had promised to let Putnam publish a number of articles, but in March, without consulting Putnam, he sent them to *Liberty Magazine*. Putnam was irritated and told Mitchell that that sort of switching was "hurtful all around." He then told Mitchell, "Only so far as this organization of mine is concerned, I have got to keep it on a definite business basis." Putnam turned down Mitchell's suggestion of revising *Winged Defense* because "the sales . . . [have] pretty well petered out."[37] From that point on Putnam was wary of Billy Mitchell's schemes. Mitchell apologized to Putnam but then pushed the editor on overseas sales of the book, hinting that perhaps the publishing house was not doing enough.[38]

Mitchell had also developed an exaggerated view of himself as a public figure and the price he could command for articles and appearances. It was also a little difficult for his supporters to understand his being almost completely oblivious to the hearings and debates in Congress over what would become the Army Air Corps Act, passed in early July 1926. While Billy Mitchell was going around the country and lecturing to those who wished to pay the $3 to hear him (and in many places they were obviously not so willing), Elizabeth continued to work on Boxwood while she tended to the child Lucy. She continued to live in Washington at 1632 19th Street, North West, while contractors gutted and restructured Boxwood, a long and costly process, and also added another wing to the house. To rebuild the fields and lawns Mitchell entered into an agreement with Issac M. Waddell, a local farmer-contractor, who would plant and harvest crops on a fifty-fifty basis. Waddell also provided nearly $1,000 worth of hay, feed, and other necessitates for the upkeep of the horses.[39] The contract did not work, for Waddell found out, as others would, that Mitchell was very slow in paying his bills.[40] At the same time, however, Mitchell was ordering three pairs of fine hunting breeches from a London tailor.[41] Betty Mitchell tended to the professional spraying and pruning of some old and quite large apple, cherry, walnut, and oak trees, as well as the English boxwoods for which the estate was named.[42] Frustration piled on top of frustrations as the state of Virginia required that a permit be granted to run electric power to the house, and the check from Betty's contractor for the permission was lost in the mail.[43]

To complicate Betty's life even more, Mitchell made plans and agreements without consulting her. Betty had to deal with a tentative agreement her husband had made to employ a sergeant named John McGuire who was on active duty in the Nashville area and planned to retire from the army in October 1926. His expertise was in handling, training, and grooming horses. McGuire had a wife and two young boys, and he wanted to begin advertising his small house for sale. His expected salary was to be $85 a month for working five days per week. Betty had to write to him, "but *I do not* [her emphasis] know just what definite plans he [Mitchell] has made about our place in Virginia. . . . For the present, there is nothing that I can do, for we will not be moved into the place until about the middle of March anyway, as I do not want to take my tiny baby out to the country in the bad weather."[44]

With his income seriously reduced and the lecture tour not earning him as much as hoped, Mitchell seemed out of touch with reality. In 1925 he bought a Packard Sport Eight automobile from an old friend, Colonel J. G. Vincent of the Packard Company. A year later, during the tour, he wrote again to Vincent, telling him that he wanted to trade the car in for a newer model, a club sedan. Vincent was not impressed by such an early trade-in on a year-old automobile the quality of a Packard; he told Mitchell to hold on to it for a while longer.[45] Mitchell was also sending a stable in Manassas, Virginia, $50 a month to board three colts. He was in hopes that Betty had contacted them about moving the colts to Boxwood.[46] The problem with that plan, which apparently had not been coordinated with his wife, was that there was no electricity or heat at Boxwood and Betty was still residing at the 19th Street address in Washington.

In April, as the tour wound down, Mitchell's name was in the newspapers again, but this time it did not reflect well on him. The *New York Times* learned that Mitchell had flown thirty-seven "flights" at Bolling Field on January 16, 1926, in a little over an hour. Flying in a new Douglas-manufactured observation plane, Mitchell took off, flew around the airfield once or twice, landed, and took off again. He was getting in the required flights to qualify for three months' flight pay, a sum of $1,027. "Accordingly he and his friends believe that he has the laugh on the army air service, inasmuch as Controller General McCarl has approved the payment. By making these flights in one day Colonel Mitchell was able to claim more than $1,000 for little more than an hour's work," the paper reported.[47] The tone of the article indicated that the writer was not amused, as this sum would have been considered handsome by anyone's consideration in 1926. When this newspaper article appeared, the Army Air Corps Act was winding its way through Congress with passage assured. President Calvin Coolidge was certain to sign this major step forward in airpower. Hugh A. Drum, Mitchell's old nemesis, appeared on Capitol Hill to propose a law that would create a national council of defense with great powers. Secretary of War Davis had encouraged General Drum to submit the legislation, which would bring together the secretaries of war, navy, and commerce and would include representatives of the army and navy. This council would assign the responsibilities for control of the land, sea, and the air.[48] Missing in all of this important debate was Billy Mitchell.

Mitchell returned from his nationwide tour in early May an exhausted man. He checked into a hospital for a few days to recover from what had to be a great drain on his body and his mind. The family was now able to occupy Boxwood, the old structure now refurbished and furnished with electricity and running water. The horses were there, and Mitchell planned to build a fine herd and recapture some of the glory days as a national horseman that he had enjoyed in 1921–1923. He had earned a little less than $40,000 on his tour, but coupled with his various writing projects his income, if wisely handled, would be sufficient.[49] Of course, Betty Mitchell had income of her own to contribute to the upkeep of Boxwood. The Mitchells chose to postpone their summer hunting trip to Africa, preferring to visit Europe the following year.

Throughout the remainder of 1926 Mitchell pushed his writing with various degrees of success. He offered an article on tiger hunting to *Liberty* magazine for $2,000, which *Liberty*'s editors rejected. Ronald Millar, the executive editor of *Liberty,* gave Mitchell some sound advice in his letter when he said, "If you were writing on some aspect of aviation or war history, and where your disclosures or opinions were of an unusually important or sensational character, then we would naturally expect to pay a larger sum."[50] The problem for Mitchell was that he had basically said all that he had to say, and many of his manuscripts were rehashes of earlier works or were highly imaginative or speculative as to the development of air and air travel. Out of Washington and out of the office of the chief of the Air Service, his sources of information had simply dried up. Mitchell also never seemed to grasp the significance of the Army Air Corps Act, and he quickly lost touch with research and development in aircraft. Mason

Patrick, who had protected Mitchell for several years, summed up Billy Mitchell by saying, "Shortly [after the court-martial verdict] Colonel Mitchell resigned his commission as an officer of the Army and since then has been but little in evidence as far as air matters are concerned."[51]

The old Billy Mitchell did surface briefly while on a hunting and fishing trip to Maine in August when he was told that Commander Richard Byrd, known for his North Pole flight in May 1926, was being pushed to become the president of the National Aeronautical Association. "It is just an attempt on the part of the navy to keep control of aviation and to retard its development so as to continue building battleships," he replied. Rather than see the navy gain control of the organization that was lobbying for airpower, Mitchell announced that he would accept the position of president.[52] Mitchell mounted a campaign to be made president and was supported by Frank A. Tichenor, editor of the *Aero Digest* and president of the large New York chapter of the association. Mitchell wrote to Tichenor, "What I want back of me is an executive board of prominent Americans [whose] statements will have weight and authority."[53] To have Mitchell as the president of such a body was certainly not welcomed by everyone. The Aeronautical Chamber of Commerce and spokesmen for the aircraft industry announced that they opposed Mitchell's candidacy. Frankly, they regarded Mitchell as a loose canon, an irritant, and a self-promoter. When asked about the stinging denunciation of his candidacy he said, in typical Mitchell style, "That doesn't bother me." During a speech in Washington Mitchell told his audience, "If we can rehabilitate the National Aeronautical Association, we will. If not, we will get another one."[54]

This was simply too much for many in the association. Frank Tichenor fueled the fires when he announced that the navy had issued "unofficial orders" to navy members to vote against the Mitchell candidacy. With so many now opposed to having the divisive Mitchell as president, Billy Mitchell withdrew from the race, saving the association from a bitter, and potentially mortal, fight.[55] A compromise candidate, Porter H. Adams, was proposed; most delegates were relieved, and he was elected. In October, Mitchell led the Washington, D.C., delegation of the American Legion to its annual convention in Philadelphia. The American Legion had always been among Billy Mitchell's strongest backers, and it was evident that he had many friends at the meeting. The Legion named a committee of nine men, including Mitchell, to study the course of American airpower and report back to the Legion at its 1927 convention in San Antonio, Texas.[56]

In November 1926, Mason Patrick, nearing the end of his military career, filed his annual report, which called for "annual air maneuvers, a separate budget for military aviation, and a separate promotion list for army aviators." He also cited the need to upgrade equipment for the Reserve Officers' Training Corps, the National Guard, and the organized air reserve. But Mitchell was still on his mind when he wrote, "When legislation pending at the close of the fiscal year is enacted, differences of opinion which have been so pronounced in the past should no longer be in evidence and all who are interested in the proper development of aviation, both commercial and military, should work together wholeheartedly and loyally to put this legislation into effect."[57] The *New York Times* and other newspapers backed Patrick, stating that though Congress had passed important legislation, it failed to appropriate funds to make a five-year air corps program work. Officers were leaving the Air Service simply because commercial companies could offer them more, and they had to consider the well-being of their families. "To serve under such a commander as General Patrick is a distinction. But officers and their families have to live."[58]

That Christmas the Mitchells celebrated their first holiday in Boxwood. The year had been filled with turmoil for everyone associated with airpower, but Mitchell was out of uniform and away from the centers of policy and power. He probably never realized that with the court-martial, the conviction, and resignation he would be so far removed. The speaking tour, which had been hectic and debilitating, was a mixed bag as far as Mitchell was concerned. The year ended on a sour note with the bitter words over Commander Byrd's possible nomination for president of the National Aeronautical Association. The rejection of a Mitchell candidacy by important segments of the air community had to come as a shock to his ego.

Perhaps 1927 would be a better year. With the upcoming presidential election in 1928, politics would certainly begin to heat up, and Mitchell was determined to see that the Democratic Party adopted his views on a unified air service and a Department of Defense. A trip to Europe to investigate airpower there would certainly be beneficial. He knew his friends in the press would buy his articles about what the former allies, the Germans, and even the Russians were doing to concentrate their air policies. In his correspondence, Billy Mitchell had never dropped the rank of general, and maybe, just maybe, 1927 would be the year to show those other generals of the Air Service and the War Department just how right he had been and how wrong they were.

NOTES

1. *New York Times*, 26 November 1925, 1.
2. Ibid., 27 November 1925, 16.

3. Foulois to Lieutenant Colonel Harry B. Jordan, 5 December 1925, in the Benjamin Foulois Papers, Library of Congress, Washington, DC, Carton 5.

4. *New York Times*, 18 December 1925, 2.

5. Ibid., 3 December 1925, 24.

6. Mitchell to Rockenbach, 14 January 1926, in the William Mitchell Papers, Library of Congress, Washington, DC, Carton 11. (Hereafter cited as the Mitchell Papers.)

7. Major E. R. Householder, Adjutant General's Office, to Mitchell, 15 January 1926, Mitchell Papers, Carton 11.

8. *New York Times*, 20 January 1926, 17.

9. Hull to Coolidge, 23 January 1926, in National Archives, Archives II, College Park, MD, Records Group 153, Records of the Judge Advocate General, General Courts Martial. William Mitchell Case Records, 1925, Case Number 168771, Entry 40, Carton 9214-5.

10. Ibid.

11. Coolidge's Written Statement, 25 January 1925, RG 153, Carton 9214-5.

12. *New York Times*, 26 January 1926, 1, 4.

13. Statement by Reid, 25 January 1926, Mitchell Papers, Carton 38.

14. Adjutant General's Office, Report of Resignation, 1 February 1926, Mitchell Papers, Carton 38.

15. Statement by Mitchell, 1 February 1926, Mitchell Papers, Carton 38.

16. *New York Times*, 1 February 1926, 5.

17. Ibid., 2 February 1926, 29.

18. Article ca. mid-February in The Hardie Collection, Golda Meir Memorial Library Archives, University of Wisconsin, Milwaukee, Carton 11, Scrapbook 1.

19. Broadside from American Legion Post No. 24, Chicago, Mitchell Papers, Carton 11.

20. Putnam to Mitchell, 28 January 1926, Mitchell Papers, Carton 12.

21. C. F. Spalding to Mitchell, 18 February 1926, Mitchell Papers, Carton 12.

22. *New York Times*, 21 January 1926, 25.

23. Ibid., 29 January 1926, 23.

24. Press Release, U.S. Air Force Association, 15 February 1926, Mitchell Papers, Carton 34.

25. Statement by Patrick, ca. Late February, 1926, Mitchell Papers, Carton 34.

26. Mason Patrick, *The United States in the Air* (Garden City, NJ: Doubleday, Doran, and Co., 1928), 182–191.

27. *New York Times*, 6 February 1926, 7.

28. Press Release, U.S. Air Force Association, 17 February 1926, Mitchell Papers, Carton 34.

29. *New York Times*, 10 February 1926, 13.

30. Ibid., 12 February 1926, 18.

31. Mitchell to Miller, ca. 15 April 1926, Mitchell Papers, Carton 13.

32. Mitchell to His Sister, 15 April 1926, Mitchell Papers, Carton 13.

33. Mitchell to Patterson, 17 March 1926, Mitchell Papers, Carton 13.

34. *The Outlook* 142 (10 February 1926), 196, 198.

35. A number of letters from Mitchell to prominent entertainment industry personalities are contained in Carton 12 of his papers.

36. Mitchell to Putnam, 15 April 1926, Mitchell Papers, Carton 13.

37. Putnam to Mitchell, 22 April 1926, Mitchell Papers, Carton 13.

38. Mitchell to Putnam, 30 April 1926, Mitchell Papers, Carton 13.

39. Contracts between Mitchell and Waddell, June to July 1926, are found in Mitchell Papers, Carton 11.

40. Waddell to Mitchell, 16 July 1926, Mitchell Papers, Carton 11.

41. H. Huntsman and Sons, London, to Mitchell, 24 March 1926, Mitchell Papers, Carton 13.

42. Elizabeth to Mr. W. B. Chenworth, 17 February 1926, Mitchell Papers, Carton 12.

43. L. G. Collier to Elizabeth, 18 February 1926, Mitchell Papers, Carton 12.

44. Elizabeth to McGuire, 25 February 1926, Mitchell Papers, Carton 12.

45. Mitchell to Vincent, 18 March 1926 and 5 April 1926, Mitchell Papers, Carton 13.

46. Mitchell to J. Carl Kincheloe, 22 March 1926, Mitchell Papers, Carton 13.

47. *New York Times*, 3 April 1926, 1.

48. Ibid., 4 April 1926, 20.

49. Burke Davis, *The Billy Mitchell Affair* (New York: Random House, 1967), 337.

50. Millar to Mitchell, 8 February 1927, Mitchell Papers, Carton 13.

51. Patrick, *United States in the Air*, 182.

52. *New York Times*, 13 August 1926, 2.

53. Ibid., 27 August 1926, 17.

54. Ibid., 28 August 1926, 10.

55. Ibid., 9 September 1926, 3.

56. Ibid., 15 October 1926, 16.

57. Ibid., 25 November 1926, 6.

58. Ibid., 26 November 1926, 18.

TWELVE

POLITICS AND DECLINE

B ILLY MITCHELL GREETED 1927 WITH A CHAMPAGNE TOAST.
Being adamantly opposed to Prohibition, he saw no need to honor such
a foolish and destructive law and had become more outspoken against it. As
Mitchell viewed the upcoming presidential race in 1928 he had decided that
the Democratic Party offered the best chance to repeal Prohibition. As a
prominent citizen of Middleburg, Virginia, Mitchell happily joined local
citizens in finding ways around the law. Cases of beer, often bought in
Pennsylvania, were brought into the county and distributed to the thirsty,
especially to members of Mitchell's polo club. One evening a smuggler of
the forbidden brew, a county doctor, was stopped by a Virginia state trooper
who was not exactly sure how to proceed against a prominent physician
when the Middleburg authorities were not too interested in the crime. The
doctor suggested that the whole problem be presented to General Billy
Mitchell for his advice. Mitchell surveyed the beer—some of it belonged to
him—and suggested that the doctor pay a "possession fine," which the worthy healer did, keeping the beer and later making proper distribution.[1]

Billy and Betty Mitchell were quickly becoming members of the local
society because of their ownership and first-rate renovation of Boxwood
and their love of horses. In late 1926 Mitchell flew over the town of
Middleburg, taking photographs for the mayor and council to illustrate possible development and conservation. Betty Mitchell interested herself in
local education matters, and in a year she was elected to the Hill School
Corporation of Middleburg.[2] With the trial, conviction, and resignation
fresh in everyone's minds, the Mitchells became local celebrities, and one
enterprising citizen printed tourist postcards showing Boxwood, the "Home
of General Billy Mitchell." The couple were convivial, and Boxwood made
a fine setting for dinners. But how long could Billy Mitchell remain content

as the country gentleman, a "foxhunter," as his neighbors called the horsey set?

Hoping that 1927 would be a productive year as far as his writing was concerned, Mitchell canvassed his friends in the publishing business. *The New York American* promised to pay $300 for four short articles, and that was a start, at least.[3] In May the *Annals of the American Academy of Political and Social Sciences* published an article entitled "Airplanes in National Defense," which turned out to be a disappointing rehash of what Mitchell had been saying and writing for several years. He again attacked the surface, or battleship, navy saying that, "On the sea the surface vessels are entirely at the mercy of air forces with the airplane. The greatest battleship is even more vulnerable than the smallest torpedo boat." He praised the submarine as the navy vessel of the future, and then returned to a familiar theme: "Air craft can stand off a hundred or more miles and launch air torpedoes carrying hundreds of pounds of gas, explosive or fire-making compounds and hit a place like New York practically every time. The air torpedo is like an airplane controlled by gyroscopes which will pilot it in the direction and for the distance desired." Ignoring recent experimentation in anti-aircraft defenses and the aircraft carrier, Mitchell called for the development of "air power, air bases and aircraft."[4] This was the essence of Mitchell the soothsayer, not prophet, and his message had not changed as technology, to which he seemed to be oblivious, marched on.

With a national presidential election only a year away, writers concerned with the national defense turned out volumes. When the very respected combat veteran Brigadier General Henry J. Riley, who had commanded an infantry brigade alongside Douglas MacArthur in the hard-fighting Rainbow Division of the Great War, wrote about national defense, his articles were cited as important to the upcoming debate. Though he recognized the coming of the aircraft carrier as vital to Pacific defenses and advocated a massive increase in air capabilities, he never mentioned Billy Mitchell as a positive influence on the expansion of airpower or on the upcoming defense discussion.[5] Articles in major newspapers such as the *New York Times* and in magazines focused on Mason Patrick, not Mitchell. Billy Mitchell was in real danger of becoming only a reference in an index, not a key player in airpower or the national defense debate.

During the spring maneuvers held at Fort Sam Houston, Texas, Mitchell's last assignment before his court-martial, the army's rejection of what Mitchell had argued for was clear. Observers were impressed with the coordination of ground and air forces. Newspaper reporters were constantly told that the air component gave to the ground commanders an opportunity to mass aircraft and strike the enemy swiftly. From Kelly Field near San

Antonio came 202 different types of aircraft to participate in the operations, which Mason Patrick and Major General Ernest Hinds, the corps commander, pronounced a great success. During the critique of the maneuvers, Patrick's assistant, Brigadier General James E. Fechet, warned ground commanders against using the aviation arm on anything but the most important missions. Hinds restated that the infantry was still the final arbiter of victory on the battlefield but that air added a new and powerful dimension to the arsenal of the commander. Secretary of War Davis ended the maneuvers by telling the press and assembled officers, "The spirit of cooperation between ground and air officers was magnificent. If there is any of that opposition of which there was evidence a year or so ago—well, if there is any officer who still has that spirit he had better get out of this man's army."[6] As far as Davis was concerned the Billy Mitchell era was over, and the army and its new air corps could proceed without loud interruptions to become an integrated combat force which was traditional yet modern. The battlefield had three dimensions, not just the air as Mitchell continued to argue.

Mitchell still received invitations to speak to various groups, but he needed some national exposure again. His thoughts turned to his typewriter and a new book on airpower. Drawing on his longtime relationship with the Hearst publishing firm, Mitchell proposed a new book on aviation, with a prepared title of *American Air Power in the Pacific*, which would be "the result of some thirty years' observation in that part of the world." As he described the proposed book to Bradford Merrill, general manager of the Hearst papers, he would cover airpower in China, Japan, India, Singapore, Java and the Dutch East Indies, and Siam. Mitchell went on to say that he would include chapters on the strategic importance of the Pacific area, the defenses of the Hawaiian Islands, the Philippines, and Alaska. Though the Hearst company was not particularly interested in the book, possibly because Mitchell had not been in the area for five years, Mitchell did have good insights as to what could very well happen if war came to the region.[7] His thoughts about the strategic importance of Alaska to both Japan and to the United States were particularly cogent.

Mitchell probably knew that the General Staff had recently finished reviewing the report he had given to Mason Patrick upon returning from his Far East inspection trip in 1923–1924. This was a full, detailed report of over three hundred pages, much of it speculative, but a great deal of it on target as far as war in that region was concerned. Since there were no storm warnings in the Pacific region at that time, the staff had approached the report as they would any document, giving responsibility to the G-2 and G-3 sections to analyze it. Also, by 1925 Mitchell was embroiled in the vari-

ous controversies that would bring about his court-martial, and the General
Staff approached anything to do with Mitchell with caution and a dose of
skepticism. As the G-3 section commented, Mitchell's view of strategy was
based on a "fundamental contention that air power has entirely changed all
previous theories of warfare and is now by far the dominating instrument
for making war."

Much of Mitchell's report dealt with the possibility of a conflict with
Japan, which he called "the vanguard of the Asiatic."[8] Billy Mitchell had
long maintained an interest in Japan and believed that a conflict between
the United States and Japan was unavoidable. Probably Mitchell had the
best view of Japan of anyone in the War Department starting in 1921.
Mitchell had an ally in Major Arthur Christie, who was in the office of the
military attaché in Tokyo and who sent Mitchell personal, lengthy reports.
"They [the Japanese] realize the necessity of an air force and are doing all
they can to come up to date. . . . They have some pretty poor stuff 'palmed
off' on them at various times by the Italians and French. . . . Both the army
and navy are expecting to ask for a large appropriation for bombing units
and personally I think they will get it. . . . The Japanese are trying to build
planes and motors, and I believe it will be only a question of time before
they can turn out a decent number," Christie wrote Mitchell in October
1921. Many Westerners doubted that the Japanese could fly aircraft in combat, but Mitchell heard from Christie that "from what I can gather I believe
they can fly well enough to make some trouble—that is if they have the
equipment and the training."[9]

After Mitchell filed his report and the staff studied it, the G-2 section
disputed his claims of increased Japanese aircraft production and the quality of that product. Christie had the opportunity to see Japanese military
maneuvers in November 1921, and he believed that the government staged
the operation more for public morale than for actual tactical practice. But
Christie felt that there was more behind the scenes than he was able to see.
The entire travel budget for the office of the military attaché was a paltry
$175, and Christie had to fund trips out of his own pocket. He did, how-
ever, receive permission to visit an Imperial Army flight school at
Tokorozawa, and he came away impressed at the potential of Japanese air.
"One thing is certain," Christie warned Mitchell, "and that is that the
Japanese are doing everything possible to improve their aviation equipment
and training and that they realize its vital importance."[10] Of great concern
to Christie and to Mitchell was a growing tendency on the part of the
Japanese to keep foreigners away from their training facilities and production centers. The G-2 section of the General Staff did not have the benefit
of the Christie observations, and the intelligence officers' dismissal of

Mitchell's warnings on aircraft production was a grave error, a mistake the United States would pay for at a later date, years after Mitchell was in the grave.

In 1922 Arthur Christie asked to be reassigned due to the high cost of living in Japan, but before he left he told Mitchell that there was almost no discussion of a separate air force in Japan because both army and navy were opposed to it, and it appeared that they would develop along different lines.[11] While Mitchell had no luck in visiting Japan to view firsthand its airpower development in 1924, he had a good picture of Japan's potential. In fact, he had a better feel for it than anyone in the G-2 or G-3 sections tasked to analyze his report. Mitchell believed, as did many officers, that difficulties with Japan over control of the Pacific would occur and that war was likely at some future date, but he failed to see the development of the Japanese aircraft carrier as a potentially deadly weapon. When he filed his report he stated that a Japanese attack against Hawaii would come after the Japanese had seized Midway Island. They would hit Hawaii, aiming at the island of Oahu, with the major attack coming against Pearl Harbor and the facilities on Ford Island. Mitchell predicted that an air attack would begin at 7:30 a.m. (the actual attack on December 7, 1941, began at 7:55 A.M. and was aimed at the navy facilities at Pearl Harbor and the air facilities at Ford Island). Mitchell, who was appalled at the lack of a real air presence in the Philippines, correctly predicted the method of Japanese attack there as well.

When the War Plans section of the G-3 reviewed Mitchell's report his comments were recorded as a rejoinder: "Since he [Mitchell] so notoriously overestimates what can be done with air power by the United States, it is not improbable that he has likewise grossly overestimated what Japan could do and would be able to accomplish with air power."[12] Mitchell's report was based on his longstanding interest in Japan's military potential, and while many of his assumptions were proven incorrect, the foundation of his report was on the mark. The problem, and the tragedy, with the report was simply this: it was Mitchell who wrote it. He had by that time become a very loud and aggravating proponent of independent airpower in the national defense, and he had caused great problems with his feud with General Menoher.

The proposed book came to nothing, but Mitchell had another possibility to offer Hearst. Mitchell cited his upcoming trip to visit Europe as a source for articles. "I have an entree into all of the countries of western Europe such as few men possess, on account of the fact that during the war I either commanded air forces belonging to those countries, fought against them, or helped others, such as Spain, in organizing their air power," he wrote.[13] Mitchell went on to tell the managing editor of the Hearst papers,

"It occurs to me that we might make some permanent arrangement about my writing for your papers on these matters [aviation questions] that is, to write you a couple of letters a week on aviation and national defense matters, or act as sort of national defense editor for your publications." As an afterthought Mitchell asked if the Hearst publications had failed to remember that he was to be reimbursed for several articles he had written sometime earlier.[14] This recalled a Mitchell in need of a forum, recognition, and income.

Mitchell and Betty departed for Paris (Lucy was left in the care of Betty's parents in Michigan), and Mitchell attended the meeting of the Paris branch of the American Legion, where there was still a tremendous amount of support for Mitchell and the unified air service.[15] Mitchell also visited Italy, Germany, England, France, and Russia. Of all the nations he visited, Italy had the greatest influence on Mitchell. There he was greeted as he had been in 1921, and he met with Italo Balbo, who would become fascist Italy's first minister of aviation, a unified air service. It was while there that Mitchell became aware of the writings of the influential Guilio Douhet, an airpower thinker.

Douhet had been involved in air matters during the Great War, and after 1917 he became head of the Italian Air Service. In 1921, after he had left the service, he published his landmark *Il dominio dell'aria*. Reflecting the thoughts of the great German military thinker Karl von Clausewitz, who believed that attacking the enemy's homeland would break the will of the people to resist in time of war, Douhet, writing almost a century later, argued that the best method for this was to use the airplane in constant attack against "vital centers." Douhet, like Mitchell, had come to believe that traditional armies and navies could no longer be the final arbiters of victory. Only the air could visit upon the enemy's homeland the horrors of modern war and could break the will of the people to support their governments and militaries.[16] Mitchell felt a kinship between Douhet's beliefs and his own thoughts on the face of the next war, and he left Italy with new ideas, but no new political ideology. He was impressed with the interest in Italy in the development of airpower, but not in Italian fascism under Il Duce. Politically corrupted Billy Mitchell was not—his political thoughts were not that deep or penetrating.

When Mitchell visited Germany, his presence caused some concern in the War Department. The military attaché in Berlin was Colonel Arthur L. Conger, who had served as Dennis Nolan's deputy in the G-2 section of Pershing's AEF staff. He brought to his post a trained intelligence eye for information and for analysis. Conger had been informed by the War Department that Billy Mitchell and his wife were traveling as private citizens and should be ignored. But Conger had been in Berlin long enough to get a sense of what the Germans were thinking, and so he put Mitchell in touch with leading German airpower advocates and industrialists. Explaining his position to Washington, Conger said that the Europeans viewed Mitchell as a leading aeronautical expert and that failing to introduce Mitchell would seem odd, if not offensive, to the Germans.[17] The 1927 visit to Europe had none of the excitement that the 1922 trip had produced, and Billy Mitchell saw very little that was new. He visited old colleagues and friends, but now that Mitchell was a resigned officer, far from the centers of power in Washington, his reception was tempered. Europe's power brokers were more concerned with who would be Mason Patrick's successor now that the chief of the American Air Service had announced his intention to retire at the end of 1927.

Then Lindbergh changed everything. On May 21, 1927, Charles A. Lindbergh, flying a monoplane designed by the Ryan Company of San Diego, California, and named the *Spirit of St. Louis,* landed at Le Bourget airfield near Paris. This son of a reformist Republican congressman who had bitterly opposed the nation's entry into the Great War had flown nonstop from Roosevelt Field, near New York, to Paris in thirty-three and a half hours. To Patrick and other airpower advocates the Lindbergh flight and the public adulation that followed his return to the United States opened up great possibilities, particularly in commercial aviation.[18] Certainly the legendary flight focused the attention of the public on private means of aviation, aviation without government subsidies and interference. Mitchell publicly praised the flight and the great achievement, but under the surface there was a feeling that Lindbergh's transcontinental flight hurt the possibilities of obtaining a unified air service in the government, since all attention would now be focused on private industry. Many of Mitchell's friends within the Air Service felt that same way. Major Ira Eaker, for example, felt that the Lindbergh flight "knocked [out] any interest" in a military flight that was planned to Latin America.[19] Lindbergh's return to the United States was triumphant, and he toured America as much as the government wanted him to.

Mitchell could not resist getting into the headlines when Charles A. Lindbergh's *Spirit of St. Louis* developed engine problems at Bolling Field, near Washington. The plane was serviced at the naval air station at the field but still could not take off. Mitchell assailed the navy for being incompetent: "Today the navy lives on principally hot air, manufactured and spread by their Washington lobby. Since the naval vessels were sunk by airplanes in 1921 and it was proved to the world that battleships were useless, the naval propagandists still hedged behind the cry that airplanes could not fly across the seas." Admiral William A. Moffett, the old Mitchell antagonist from naval air, was restrained from answering Mitchell, but Acting Secretary of the Navy Theodore Douglas Robinson responded, "What's the use of getting into an argument with a man who, on the face of it, doesn't know what he's talking about and, if he did, couldn't tell the truth about it anyway."[20] Lindbergh was embarrassed by the whole mess, and he issued a statement complimenting the navy, saying that "it in no possible way could have been caused by carelessness on anyone's part but developed as do most minor troubles." In the War Department, Davis, Patrick, and Fechet probably breathed a sigh of relief that Mitchell was no longer in uniform, and it did appear that Mitchell's statement, which ended with the usual call for a unified air service, was geared to take unfair advantage of a minor situation.

The Mitchells returned to Boxwood, and Billy began the process of writing articles for various magazines and newspapers. While he was sitting at his typewriter, changes were being wrought at the Air Service in Washington. The old Air Service, born in the Great War, was about to become a name of the past. Mason Patrick was to step down after forty-five years of service. His final report, filed before Thanksgiving 1927, summarized his six-year association with the service, and he called, to no one's surprise, for increased funding for new planes and for training. Like many of his colleagues in aviation, Patrick was quick to cite the advances in long-distance flying by the military while downplaying the significance of the Lindbergh flight.[21] Editorials praising Patrick appeared in almost every major newspaper, and none was more complimentary than the *New York Times,* which said in part, "On military aviation there has been no better authority than General Patrick. No army officer has ever done more to advance commercial flying in this country."[22] Very few major newspapers even mentioned the name of Billy Mitchell.

Brigadier James E. Fechet was named head of the Army Air Corps with a promotion to major general, and Mitchell's old enemy, now a brigadier general, Benjamin D. Foulois was slated to be the new assistant to Fechet. Both Fechet and Foulois could be counted on to maintain the steady approach established by Patrick and to slowly move toward a separate air arm with a minimum of fuss and headlines; there would be no repeats of the Billy Mitchell fandango of 1925 with those two. Younger officers understood that these two senior aviators were not to be trifled with, nor would they allow any Mitchell supporters to make headlines. Dealing with Mitchell and the upcoming presidential election was another matter indeed.

With Christmas 1927 and New Year's Day 1928 over, Betty and Billy Mitchell needed to spend some time at Boxwood. Mitchell had financial problems and had negotiated a loan from his aged uncle Washington Becker, who was living in Milwaukee. After two weeks of pencil and paper drills, Mitchell reckoned that his outlay for 1927 was $56,221.00, while his income was $53,756.00. He had lost about $700 on the stock market during the year. On the bright side, he noted that his writings had paid for his European trip and that income from Boxwood equaled expenditures for the farm. However, Betty and Billy Mitchell figured that they owed the government about $9,000 in taxes.[23]

On January 22, 1928, there was a fire at Boxwood that threatened the structure and quite possibly the lives of Billy and Betty Mitchell. An old stone fireplace in the master bedroom had broken down over the century of its existence, and it caught some timbers on fire. The couple smelled the fire, located it, and with fire extinguishers and buckets of water from the bathroom ended the flames, but not before it did about $1,500 in damage.[24] The next day Mitchell wrote to Sidney Miller to tell him about the fire and damage, but half of the letter concerned a new project, the establishment of an aeronautical university in the Washington area. It seemed to be a very inappropriate time to be informing his father-in-law of future plans, when the initial news reports about the fire stated that Boxwood had burned to the ground and that the Mitchells were either homeless or worse, could not be found and were feared to be dead. But Billy Mitchell could not be restrained, sharing plans of a department of experimentation and a department of new inventions with his father-in-law.[25] It must have appeared to Sidney Miller that his son-in-law had problems with what was important and what was not.

The ever-generous father-in-law responded to Mitchell, "I have already written Betty that if the insurance money is not sufficient to put the whole place in proper shape, she should draw on me." As far as the university was concerned, Miller was less than encouraging, only providing Mitchell with the names of two individuals who might be interested in donating money. Sidney T. Miller was very careful not to pledge anything other than moral support and a few names in the Detroit area.[26] Planning an airpower think tank, and that was exactly what it was to be, was a major project, but Mitchell still needed income. In early February, *Collier's Magazine* agreed to pay him $1,000 for an article on new scientific inventions and national defense. The eager editor suggested the first sentence: "Do you know that it is possible to hit New York, or Boston, or Philadelphia with an aerial torpedo released from any point in Eastern Europe?"[27] G. D. Eaton of *Plain Talk* magazine also agreed to take an article from Mitchell.[28] With the presidential campaign coming in the fall Mitchell would have his name again before the public and before the Democratic candidate.

With much fanfare, Mitchell called a press conference in Washington, D.C., on February 8 to announce the campaign to fund the United States Aeronautical University. He stated that it would take $10 million to start what Mitchell claimed would be a nonprofit organization to teach men and women "a higher standard of efficiency in both civil and military flying." Mitchell went on to say that the university would be linked with the United States Air Force Association, which had an announced political agenda. It was the same old Mitchell line—a separate air service with a cabinet officer to see to the development of commercial and military air. During his conference, Billy Mitchell went back to very familiar and tired themes: the *Ostfriesland* sinking had proven airpower over the battleship navy; advancement in the air was stymied by bureaucrats, especially in the navy; and real invention was ignored by entrenched powers in both the War and the Navy Departments. There were hints during the presentation that this new institution would have Billy Mitchell as its head.[29] The response to the conference, even with Eddie Rickenbacker there to pledge to support the university, was less than enthusiastic, and as Mitchell turned more and more toward the elections his interest waned, as did contributions. Finally the whole project was abandoned for lack of support.

In April Mitchell used his father-in-law's connections to convince the respected *Atlantic Monthly* to accept an article. Reflecting his visit to Italy and his newfound interest in Douhet's writings, his letter to the magazine's editors indicated that "a future war . . . will consist in armies holding the ground, aircraft striking directly into vital centers of the enemy country, no matter where it may be, in Europe, America or Asia." He told the editors about the *Ostfriesland* bombings and his fight against the navy, which everyone must have heard before. But he did drop names, men with whom Sidney T. Miller had contact.[30] The resulting article appeared in September and was entitled "Building a Futile Navy." There was precious little of Douhet in the article and much concerning aircraft that were invisible from the ground, aerial torpedoes, and noiseless airplane engines.[31] Taking advantage of his recent European trip, Mitchell did prepare an article on Russian aeronautics for *The Forum* magazine, for which he received the sum of $150. His insights into European development would have been a great contribution had Mitchell not been so focused on the old battles and the upcoming election.

The thousand-dollar *Collier's* article appeared in late April and gave warnings about what could happen if New York City were bombed by the air or struck by the now talked of aerial torpedo launched from Western Europe. It would be impossible to ferry troops across the Atlantic, as the United States had done in the Great War, because aircraft and submarines "would sink the whole convoy." The poor old *Ostfriesland* was sunk again, and Mitchell stated that "contrary to popular belief, it is harder to hit a still target on the water from the air than one that is moving." That was quite interesting in that all of Mitchell's bombing targets were stationary. Toward the end of the article, and in bold type for emphasis, Mitchell told his audience, "The Naval Airplane Carrier is merely an expensive and useless luxury used principally as propaganda by the Naval Services to cover up the fact that they have NO adequate defense against aircraft." Back to the theme of a Department of National Defense with the air on an equal footing with the now useless army and navy, Mitchell went.[32] Mitchell never seemed to realize that editors were aware of what was being published in other magazines, and soon this "Johnny One Note" approach would wear thin.

One consistent source of support for Billy Mitchell was the American Legion, an organization no politician could ignore. In March Mitchell went to a planning session of the Legion's National Defense Committee with several resolutions in his pocket. He had been a long-time member of the National Defense Committee, and as such he would have a heavy impact on the positions the Legion would take. But Mitchell was surprised at the opposition within the committee to his resolution calling for a "single department of National Defense under a Secretary with Sub-secretaries for the land, sea, and air and munitions." The committee of nine former soldiers cast eight votes against, and one, Mitchell's, for. Furious at the committee, Mitchell demanded the right to submit a minority report for the convention to debate.[33] When the Legion convention opened in San Antonio in October and resolutions of the National Defense Committee were presented by retired General Hason Ely, who had commanded the 28th Infantry Regiment during the AEF's first fight at Cantigny in 1918, the convention sided with the committee and not Mitchell. After a series of motions and loud debate, the Legion did affirm Mitchell's idea of a department of national defense, and that was a small, but important, victory for Mitchell for the elections of 1928.

Going into the June Democratic Convention in Houston, Texas, it was fairly certain that four-term governor of New York, Alfred E. Smith, would be nominated to carry the party's standard in the upcoming election. What was not clear was who Smith would designate as his running-mate. Al Smith was going to be a hard sell for the country: Roman Catholic, antiprohibitionist, associated with New York City's Tammany Hall, and clearly a product of the urban East. Placing Smith's name in nomination was another New Yorker, Franklin Delano Roosevelt, and Smith breezed to acceptance by the party faithful, including Billy Mitchell. The Wisconsin delegation, which had twenty-six votes, announced that it would caucus to consider placing Mitchell's name in nomination for the position of vice president.[34] An organization called the Billy Mitchell Volunteers was formed in Washington to bring his name before the convention as a presidential candidate.[35] Nothing would come of the Billy Mitchell Volunteers effort, since Smith went to the convention as the front-runner for the nomination and needed to balance the 1928 ticket. Senator Joseph T. Robinson of Arkansas was selected to be the second man on the ticket. The Democratic Party had been out of the White House since the election of Warren G. Harding in 1920, there was an intense campaign to "throw the rascals out," and Mitchell could nonetheless play a role as a man with national name recognition.

The Houston convention was disappointing for Mitchell because there had been much speculation that Smith would promise to make Billy Mitchell secretary of war in his cabinet. But this was only a rumor; nothing substantive was offered to Mitchell.[36] The Democrats were also hesitant to make any commitment to establishing a department of national defense. Mitchell returned to Middleburg and waited until July 30 to hold a sparsely attended press conference giving his unqualified support to the Smith-Robinson ticket. "Dishonesty has been rampant everywhere. The only difference between the Harding Administration and the Coolidge Administration is that the Coolidge people covered up more. The remedy to the situation is electing an honest man like Alfred E. Smith who is a human person and knows how to deal directly with people," Mitchell proclaimed. He promised that he would campaign for Al Smith all over the country.[37] What Mitchell did not add was that he would do his stump speaking at his own expense, and that would be considerable at a time when Mitchell's finances were weak.

In the fall Mitchell moved temporarily to his uncle's home in Milwaukee to have a Midwest base for his campaigning. Billy Mitchell had taken a few weeks in August to do some duck hunting in North Dakota, with great success. He then traveled to Albany, New York, to confer with Al Smith before undertaking a speaking tour, but he received no promises about the War Department or about the establishment of a department or secretariat for air. He personally liked Al Smith, and breakfasted with him and his family.[38] Some old friends of his, however, were upset over Mitchell's support for Al Smith and his intemperate language about the Coolidge administration and the Republican candidacy of Herbert Hoover. M. Robert Guggenheim of New York, who knew Mitchell from various thoroughbred dog shows and sales, bluntly told Mitchell, "I am sorry that you are campaigning for Al Smith, as I am afraid you are barking up the wrong tree."[39] Responding to Guggenheim, Mitchell said that "even if we don't get our candidate elected this time, we will get a party together that will form a strong, liberal organization around which young Americans can rally."[40]

Mitchell remained upbeat throughout the election, working tirelessly in Wisconsin and the Midwest for Smith. He planned a statewide speaking tour in his native state the last two weeks of the campaign, while making appearances in neighboring states.[41] It is easy to gather that Mitchell was firmly convinced that a Smith victory would bring a department of defense, and in his mind there was no one better to head it than himself.[42] But all was not campaign stops and German sausages in Wisconsin for Mitchell. He was having financial difficulties because so much of his tireless campaigning came out of his own pocket. By early October he had accumulated $1,900 in overdrafts at the Riggs National Bank in Washington.[43] By selling some stocks and manipulating some funds he had, he was able to cover his overdrafts,[44] but debts were mounting. Despite his up-and-down finances, Mitchell refused to endorse Lucky Strike Cigarettes because he was a longtime opponent of tobacco use in any form.[45] He did take time to appear in Chicago as part of the Boston Store's aeronautical exposition. The downtown Chicago department store sent Mitchell a much-welcomed check for $300.[46]

Mitchell's set stump speech was well written, but he did not have a fiery delivery, and as a public speaker he left much to be desired. "Economic policies must march hand in hand with political policies, otherwise one will run counter to the other and a crash will result sooner or later. This crash may be economic, it may be political or it may be both. Where political disaster results, it makes the people unhappy and makes them lack confidence in their form of government," claimed Mitchell, echoing the party line in the election. In an agonizingly complex discussion of the differences between the conservative Republicans and the liberal Democrats, Mitchell arrived at a point where he said that the Republican Party had since the Great War allowed government to centralize power in the hands of the bureau heads and secretaries, and the states had nothing to say about it. Reflecting his years as a soldier, Mitchell went on to say, "So then we see the Republican party entrenched in its citadel, behind rows of money bags, with its artillery of propaganda and press bureaus, moving pictures, radio and even airplanes deployed all over the United States. Their banner bears the slogans 'Ignorance on the part of the people is bliss for us.'" Al Smith's election promised decentralization of power, fair money policies, and fairness in foreign policy, and he would put an end to Prohibition (Mitchell spent a third of his speech attacking it, for the benefit of beer-loving

Wisconsinites, to be sure). Yes, Smith was a Roman Catholic, but Herbert Hoover was a Quaker, and could a Quaker, whose faith lauds the conscientious objector, be capable of leading America in war? Mitchell thought not. Perhaps understanding the mood of Smith and the Democratic Party bosses, Billy Mitchell omitted any reference to a department of defense and a unified air service.[47]

Al Smith went down to a solid defeat, carrying only eight states, and Mitchell finished the campaign deep in debt. There was now no question that he was solidly in the party of grandfather and father, but Herbert Hoover was in the White House. There would be no department of defense or unified air force, and, for Billy Mitchell, there would be no job in Washington. Mitchell now had to turn back to Boxwood, his horses, his writing, and a new book that he had been planning for some time, a glimpse into the future of airpower. After the Christmas holidays Mitchell received a very large check for $14,593 from the annual income from his mother's estate.[48] This welcomed windfall would ease some of the financial problems incurred during the Al Smith campaign, but Mitchell was investing in the stock market, often without much luck. *Collier's Magazine* agreed to take an article on tiger hunting for $1,000.[49] Going into the new year of 1929 it appeared that Mitchell's finances would be on better ground.

No one could have predicted the dire economic situation by the end of the year, and the depression starting in October 1929 would seriously affect everyone, including Mitchell. He believed that he had come out of the Al Smith campaign as a major figure in the Democratic Party because of his campaigning. Believing that the next election would bring a Democrat to the White House, he thought he would indeed eventually become a cabinet member. However, a problem arose concerning where Billy Mitchell considered his primary residence to be. He maintained a Washington, D.C., address through the American Legion, and he claimed Boxwood, in Virginia, as well. But during the Smith campaign he had made much of his Milwaukee ties, telling one Wisconsin congressman that he had never given up his Wisconsin associations. "I fully intend in the future to keep up my connections with Wisconsin. I live a good deal of the time in Middleburg, Virginia. I have not decided yet to change my legal residence."[50] What Mitchell had overlooked was that he needed strong ties with a state party to have a chance at a cabinet position. Those strong associations with either the Wisconsin or Virginia Democratic Party he simply did not have. Mitchell had a national reputation and was a willing worker for the party, but he was not a real power-broker. His trial had occurred four years before, and Mitchell as national figure was fading from view. This indecision over state association in 1928 would come back to haunt

Mitchell during the next presidential election. But that was in the distant future, and the great crisis of October 1929 would change the American dynamic.

NOTES

1. Eugene M. Scheel, *The History of Middleburg and Vicinity* (Middleburg, VA: Middleburg Bicentennial Committee, 1988), 134. Copy provided by Daniel H. de Butts of the Fauquier Historical Society, Warrenton, VA.

2. Ibid., 125.

3. Memorandum found in the William Mitchell Papers, Library of Congress, Washington, DC, Carton 13. (Hereafter cited as the Mitchell Papers.)

4. William Mitchell, "Airplanes in National Defense," *Annals of the American Academy of Political and Social Sciences*, 131 (May 1927), 38–42.

5. Henry J. Riley, "Our Crumbling National Defense," *The Century Magazine* 113, 5 (March 1927), 513–522.

6. *New York Times*, 21 May 1927, 40.

7. Mitchell to Merrill, 10 June 1927, Mitchell Papers, Carton 13.

8. Mitchell's Report, December 1923, Mitchell Papers, Carton 42.

9. Christie to Mitchell, 28 October 1921, Mitchell Papers, Carton 9.

10. Ibid., 28 November 1921.

11. Christie to Mitchell, 17 June 1922, Mitchell Papers, Carton 9.

12. Cited in Burke Davis, *The Billy Mitchell Affair* (New York: Random House, 1967), 183.

13. Mitchell to Merrill, 10 June 1927, Mitchell Papers, Carton 13.

14. Ibid.

15. "That Still-Born Battle of 1926," American Legion in Paris, ca. 1927, in the Ira Eaker Papers, U.S. Army Military History Institute Archives, Carlisle Barracks, PA, Carton 2. (Hereafter cited as the Eaker Papers.)

16. Patrick Hurley, *Billy Mitchell: Crusader for Air Power* (Bloomington: Indiana University Press, 1975), 114.

17. Ibid., 115.

18. *New York Times*, 26 June 1927, Section 9, 1.

19. Eaker to Herbert Dargue, 17 June 1927, Eaker Papers, Carton 2.

20. *New York Times*, 17 June 1927, 4.

21. Ibid., 24 November 1927, 16.

22. Ibid., 16 December 1927, 24.

23. Mitchell to Washington Becker, 20 January 1928, Mitchell Papers, Carton 14.

24. Ibid., 23 January 1928.

25. Mitchell to Miller, 23 January 1928, Mitchell Papers, Carton 14.

26. Miller to Mitchell, 24 January 1928, Mitchell Papers, Carton 14.

27. William L. Cheney to Mitchell, 7 February 1928, Mitchell Papers, Carton 14.

28. G. D. Eaton to Mitchell, 11 February 1928, Mitchell Papers, Carton 14.

29. *New York Times*, 8 February 1928, 8.

30. Mitchell to *Atlantic Monthly*, 13 April 1928, Mitchell Papers, Carton 14.

31. William Mitchell, "Building a Futile Navy," *The Atlantic Monthly* 142 (28 September 1928), 408–413.

32. William Mitchell, "Look Out Below," *Collier's Magazine* 81 (21 April 1928), 8–9, 41–42.

33. "Summarized Report of the National Defense Committee," American Legion Convention, ca. March 1928, Mitchell Papers, Carton 45.

34. *New York Times*, 26 June 1928, 3.

35. Shirley Holladay to Mitchell, 25 June 1928, Mitchell Papers, Carton 14.

36. There are numerous references to the possibility of Mitchell being named to head the War Department in a Smith administration. These are found in the Mitchell Papers, Cartons 13 and 14.

37. *New York Times*, 31 July 1928, 5.

38. Mitchell to Fred Osborn, 19 September 1928, Mitchell Papers, Carton 14.

39. Guggenheim to Mitchell, 26 September 1928, Mitchell Papers, Carton 14.

40. Mitchell to Guggenheim, 28 September 1928, Mitchell Papers, Carton 14.

41. Mitchell to Henry Cheney, 29 September 1928, Mitchell Papers, Carton 14.

42. Mitchell to Harold Hartney, 29 October 1928, Mitchell Papers, Carton 14.

43. Riggs Bank to Mitchell, 3 October 1928, Mitchell Papers, Carton 14.

44. Mitchell to Riggs Bank, 30 October 1928, and Riggs Bank to Mitchell, 1 November 1928, Mitchell Papers, Carton 14.

45. John C. O'Laughlin, Editor of *The Army and Navy Journal*, to Mitchell, 30 October 1928, and Mitchell to O'Laughlin, 3 November 1928, Mitchell Papers, Carton 14.

46. S. F. Lewis to Mitchell, 25 September 1928, Mitchell Papers, Carton 14.

47. Mitchell, Speech, "My Views on Some of the Campaign Issues," Mitchell Papers, Carton 26.

48. First Wisconsin Trust Company Report to Mitchell, 29 December 1928, Mitchell Papers, Carton 26.

49. William L. Cheney to Mitchell, 18 January 1929, Mitchell Papers, Carton 26.

50. Mitchell to John C. Schafer, Representative, 4th District, Wisconsin, 24 April 1928, Mitchell Papers, Carton 26.

THIRTEEN

THE LAST FLIGHT

W HEN BILLY MITCHELL WAS AT HOME AT BOXWOOD WITH Betty and Lucy he could be an attentive husband and father. He certainly loved his horses, dogs, and hunting, and the old Virginia home was open to an interesting cast of characters. In Middleburg he was known as General Billy, and the town took to the eccentric man who had occupied headlines in 1925 and 1926. The early months of 1929 seemed to be the happiest for Mitchell in many years, and he lavished attention on his young daughter Lucy. It seemed obvious to Billy and Betty Mitchell that Lucy had eye problems, and a local doctor prescribed glasses for the child. As a four-year-old, showing a definite Mitchell trait, Lucy rebelled and refused to wear them. Billy decided, much to Betty's distress, that if Lucy did not want to wear them then she did not have to. But there was a reason for Mitchell's acquiescence. He went to the local dry good store in Middleburg and bought cheap dime-store glasses for everyone at Boxwood—cook, groom, maid, Betty, and himself. This left little Lucy as the only person without a pair of glasses, and she decided that it was high time she became a spectacle-wearing member of the Boxwood family. Over a short period of time the cheap glasses disappeared, and Lucy retained her glasses.[1]

Betty Mitchell was busy with the Hill School Corporation of Middleburg, and she was instrumental in securing a $10,000 loan to add an auditorium to the school. After organizing numerous fundraisers the Hill School parents, Betty included, added windows, chairs, and stage curtains for the auditorium, which made it one of the most modern school facilities in northern Virginia. Needing no coaxing at all, Billy Mitchell agreed to be the principal speaker at the structure's formal dedication on March 21, 1929.[2]

When he was not being a Virginia country squire attending to his horses and dogs, Mitchell was busy at his typewriter, working on several projects. The major one was a book of memoirs about his service in the Great War. The year 1928 marked the tenth anniversary of the great St. Mihiel and Meuse-Argonne battles and the November Armistice. A number of generals had already published their memories of the Great War including James Guthrie Harbord, Pershing's chief of staff; Hunter Liggett, commander of the 1st Army; and Charles Gates Dawes, recently Calvin Coolidge's vice president. John J. Pershing was rumored to be working on his own massive memoir of the war, and Peyton Conway March, the controversial army chief of staff, was supposedly preparing his own book.

What really grabbed Mitchell's attention was the publication of Mason Patrick's book *The United States in the Air* in the fall of 1928. Patrick's book was lackluster—no elegance of phrase or color of description. But it was a useful outline of Patrick's work as chief of the Air Service during and after the war, and he had much to say about Mitchell, most of it friendly. What seemed to aggravate Mitchell the most was Patrick's prediction that a department of national defense would be created with assistant secretaries for army, navy, and air.[3] These were Mitchell ideas, which he had advocated for years, and he was not going to be upstaged by Patrick. After Mitchell's conviction and resignation, Patrick and Mitchell did not speak or correspond, and at no time did Mitchell acknowledge Patrick's great role in the wartime Air Service or as his position of chief after the war.

Mitchell threw himself into the preparation of the manuscript, which was long and rambling. By April the manuscript was complete, and Mitchell sent it to the respected Boston publishing house Little, Brown, and Company, which had indicated some interest in publishing memoirs by Great War figures. In a short time, Herbert F. Jenkins, the vice president of Little, Brown, rejected the manuscript for a number of reasons, which should have been a danger signal to Mitchell. In typical Mitchell style he had urged the company to produce the book in two volumes, but Jenkins told Mitchell, "I believe the manuscript would be materially improved if intelligently edited and somewhat curtailed in parts. I believe . . . that you make a mistake in your preface [where] you state that 'much of war history is written to bolster damaged military reputations,' and . . . your attacks on the regular army and navy organization are futile. You have such a big story to tell that it is really a pity you so frequently introduce propaganda for a separate air service. . . . From a publisher's standpoint this mars the book from a sales point of view."[4] The manuscript and numerous photographs were returned to Mitchell with no request for revision and no mention of resubmission. Mitchell would instead publish a series of articles about his participation in the Great War, and they were short, better edited, and free of his continual harangues about the air service, army, and navy.

In March 1929, *National Geographic* published a Mitchell article filled with quotes such as "Battleships may become like armored knights of old."[5] Mitchell's output of paid articles was dropping rapidly as it became apparent to editors that he was only rehashing old themes and writing in generalities, using data from the 1921 *Ostfriesland* and *Alabama* tests. But it was difficult for Mitchell to realize that his glory days had come in 1925 with the committee hearings and the court-martial. With Charles Summerall as chief of staff and James E. Fechet and Benjamin Foulois in the Air Corps office, he had no friends in Washington. This isolation meant that he had little idea of the currents in the air service, and though old colleagues sent him some information, Mitchell remained outside the mainstream. Fewer and fewer letters came from men like Hap Arnold or Carl Spatz, officers who were a part of the evolving air force of the United States Army.

Life was not that unpleasant for Billy Mitchell, however, because he had Boxwood and his horses and dogs. Throughout the summer of that last year of the roaring twenties, the Mitchells traveled a good deal, going to New York and Maine. Fishing occupied much of Mitchell's time, satisfying partially his need to be outdoors and physically active. His walking stick was with him always, but friends noticed that he used it more for walking than for show. At age fifty he had flair-ups of rheumatism, and he flew less now than he formerly did. Even so, Betty became pregnant again, and the next year William Mitchell Jr. would be born. He continued to work on another book, to be published under the title *Skyways,* but sales would be very disappointing.

Mitchell lost a good bit of money in the stock market crash of October 1929. When Mitchell had extra money he had invested heavily in the stock market, and he often did not invest wisely.[6] He had never been good with handling money, but Boxwood was paid for and Betty had her own income. Like many Americans, though, the Mitchells faced some hard times adjusting to the new economic and political situation in the United States. It was not lost on Mitchell that the Depression could work a new dynamic in American politics, giving the Democratic Party a real opportunity to capture the White House in 1932. During Mitchell's tour for Al Smith the year before he had spoken in general terms about a crash of some sort, political and economic, and by the end of 1929 he turned his thoughts to the upcoming congressional races.

Politics might be interesting and perhaps rewarding for Mitchell, if the right Democrat went into office, but Mitchell's focus was on Boxwood and his writing. William Ziegler Jr. of New York wrote to Mitchell that December in his capacity as the president of the Polo Committee of the Ox Ridge Hunt Club of Darien, Connecticut. Ziegler had known Mitchell for years through horse shows and polo events, and he wanted Mitchell to help him find a manager for the club. Billy Mitchell's reputation as an accomplished horseman, hunter, and breeder was solid, and he was well connected throughout the world of thoroughbred horses. Ziegler hoped that Mitchell might know a retired army officer with possibly a cavalry background and managerial skills who could run the Darien club.[7]

Mitchell knew just the man to find someone to take the position. He was Colonel George S. Patton Jr., who was serving in the office of the chief of cavalry in Washington. If ever two persons were cut from the same bolt of cloth it was Billy Mitchell and Georgie Patton; mavericks, believers in new forms of warfare, disdainful of established authority, and outspoken almost to the point of insubordination. Their relationship began with horses and hunting, and Patton hunted with Mitchell a number of times in northern Virginia. Patton had spoken in favor of Mitchell during the controversies and trial in 1925, but they had differences over the employment of airpower on the battlefield. Mitchell wrote to Patton to ask if he knew someone who was about to retire from the army to take the job at Ox Ridge and to invite George and Beatrice Patton to visit Boxwood.[8] Patton immediately responded, suggesting two colonels of cavalry who were nearing retirement.[9] Mitchell sent Patton's recommendations to Ziegler, and he told his hunting friend from New York that Patton was the type of man one should listen to.[10]

Over the next several years Mitchell and Patton were together often at foxhunts and horse shows, and while neither man changed the other's mind about airpower they maintained a good relationship, especially when they turned their conversations to the men who ran the War Department or the General Staff. Patton's thought on airpower remained fairly traditional— using airpower to support ground operations—and Mitchell had little grasp of Patton's forte, armored or mechanized warfare. In fact, many of Mitchell's former subordinates, enthusiastic aviators like Henry "Hap" Arnold, saw the possible deadly combination on the battlefield of air and mechanized cavalry units, ranging far beyond the commander's vision to seek out and find the enemy.[11] James E. Fechet and Benjamin Foulois had been preaching air-ground coordination since Fechet took over as chief from the retired Mason Patrick. What would become known as the "integrated battlefield" with all arms working in concert, giving width, depth, and height to combat, was what Mitchell did not envision.

When Mitchell wrote his immediate postwar paper on the tactical application of airpower in 1919, he came close to the use of air in the team concept, but those ideas had long ago been replaced by his support for a unified air service and the department of national defense. Billy Mitchell had never manifested an intense interest in technological innovations and in changes in aircraft design, and often his writings reflected the fact that he was not current on new capabilities of engines, communications, and aircraft manufacturing. He returned constantly to the *Ostfriesland* bombings of 1921, but never appeared to see the great changes in aircraft carrier capabilities, for example. This would effect his capability to turn out articles in the 1930s because it was clear to editors that he really had nothing new to add to the air debate. Mitchell's speculations about the future of aerial warfare were interesting, and he wrote well about air torpedoes and the like, but as far as the tactical and strategic questions of the day were concerned he could add very little.

Over the Christmas season of 1929 the Mitchells traveled, as was their custom, to Detroit to spend the holidays with the Millers. In Detroit Billy Mitchell could see the growing effects of the Depression in the city that produced so many American automobiles. His father-in-law, a lawyer for the Packard Corporation, gave him alarming facts and figures, and this picture was reinforced by visits with Eddie Rickenbacker and others. To those who supported Herbert Hoover in 1928 Mitchell adopted an irritating "I told you so" tone. Mitchell told Sidney Miller that he had contracted with the *Saturday Evening Post* and other magazines for articles, and he had finished a new book manuscript. Financially, even with his losses in the stock market in 1929, things were looking up for him, but he had great concerns for the stability of the country in the wake of the deepening depression. His visit to Russia just a few years before had filled him with a deep hatred for the Soviet system, and though he admired their emphasis on civilian and military aircraft, Mitchell had no personal doubts about the dangers of their system. One of the things he noticed while in Russia was its leadership's ability to effectively use propaganda, in this case to support Russian air activities, and he worried that such an expertise in capturing the hearts and minds of people could make inroads in a country facing the serious economic and social woes that now beset the United States. Any hint of "Soviet influence" would send him into a rage, and it would in 1931 and 1932 influence his relationship with the American Legion and with World War I veterans.

Mitchell produced several articles for the *Saturday Evening Post* that were mild by Mitchell standards. One was a survey of world aviation in which he praised Italian efforts in unifying their air forces. He stated that because of the Alps that lie to the north of Italy, the Italians had developed aircraft that could fly over the mountains and reach France and Austria. If there were a war, Mitchell postulated, Italian aviation could very well secure control of the Mediterranean Sea. Advances in Italian aviation were mainly military, but as Mitchell saw it, the enlarging of the Italian air arm was due to the centralized Ministry of Aviation. (As in the past, Mitchell made no comment about the fascist state in his article, nor did he praise Mussolini.) When Mitchell surveyed the state of Japanese aviation he allowed that the empire's aviators were indeed fine flyers, but he totally overlooked the development of the aircraft carrier and Japanese naval aviation.[12] His other article was a lighthearted, delightful look at the sport of flying and how the airplane could help the hunter, the fisherman, and the sportsman gain access to formally inaccessible areas.[13]

In the fall of 1930 Mitchell did limited campaigning for Democratic Party candidates for the House and the Senate. The state of his finances did not allow the extensive travel that he had undertaken for the Smith campaign two years earlier, but his appearances were marked by attacks against Hoover and the Republicans over such issues as airpower, the Depression, and, of course, Prohibition, which Hoover had embraced as a "noble experiment." The fall convention of the American Legion disturbed Mitchell because he saw what he believed to be a dangerous shift toward radicalism over the question of the World War I bonuses for veterans. In 1919 the Legion had surfaced the idea of "adjusted compensation," or the difference between what a soldier earned and what he could have earned in the civilian sector had he not been called to service. In 1924 a law was passed over President Coolidge's veto that gave a soldier $1 per day if he served in the United States or $1.25 per day for service in France. The payment of the bonus was deferred until 1945. By the Great Depression, over 3.5 million of those "adjusted compensation certificates" had been issued, and in January 1931, the Legion demanded immediate payment to help former soldiers suffering from unemployment and severe economic privation. Mitchell, who always maintained the persona of an independently wealthy man, looked askance at the growing demands for payment coupled with a threat of action if the government did not act.

By 1931 Mitchell's interest in politics was heating up. In July he brought his boat Canvass Back to the docks of the Capital Yacht Club where he was a member and an elected vice commodore and hosted New York Governor Franklin Delano Roosevelt to talk about politics and the upcoming Democratic National Convention in 1932. In a letter to an old Wisconsin friend, Mitchell said that he was greatly impressed with Roosevelt and would support him for president.[14] Roosevelt had opposed

Mitchell's calls for a unified air service in 1919 and the early 1920s, but now the New Yorker called Mitchell "Dear Billy." Mitchell also took time to apply for his $35 per month Spanish-American War pension.[15] From the meeting with Roosevelt on, Mitchell's attention was consumed by the presidential race. He believed with good reason that the Democratic Party with Roosevelt at its head could win the presidency in 1932, and it stood to reason that Mitchell could head a new air secretariat or a new department of defense.

One of Mitchell's friends was Arthur Brisbane of the *New York Evening Journal,* and Mitchell used him as a sounding board for his political ideas. Mitchell was gravely concerned about Japanese action in Manchuria. "The aims, mentality and psychology of the Japanese are understood by very few people in this country. Some day we will have an armed contest with them, as we are the only great white power [whose] shores the waters of the Pacific touch."[16] Brisbane and Mitchell had met through their mutual interest in horses, hunting, and politics. It was through Brisbane that Mitchell made contact with other men who were influential in the publishing business. T. V. Rank of the Hearst newspapers was one of those Brisbane introduced Mitchell to during one of Billy's trips to New York in early 1932. Rank agreed to publish several articles "based on the premise that Japan is actively preparing for a conflict with the United States." Months passed before Rank received a poorly written article entitled "Suppose Japan Surprised Us," and Rank was furious, giving Mitchell a very pointed dressing down. The article gave no background material, nor did it indicate when or how the Japanese would attack the United States.[17] When Mitchell responded to Rank he stated, "I have been very busy with politics. It looks to me now as if Governor Roosevelt will certainly be elected and come pretty close to having a landslide."[18] It looked that way to Rank too, but campaigning did not justify treating an editorial manager of the Hearst newspaper chain in such a cavalier manner with such a poorly thought-out written article.

By 1932 Mitchell had made his mind up to be as active in the current campaign as he had been for Al Smith; he probably had been encouraged to do so with a hint that he could fill a post in a Roosevelt administration;[19] of course, Mitchell was lobbying for a position. In May 1932, he had taken a position against the Bonus Marchers, a group of war veterans who converged on Washington, writing, "The [veterans] will be prey to Communistic propaganda and with their semi-military organization, if they obtain arms, they may attempt to seize the reins of government."[20] Though this was unlikely, Mitchell's attitude toward anything that smacked of the propaganda he had seen firsthand in Russia was harsh, even to the point of turn-

ing a blind eye to the sufferings of his old comrades-in-arms. He publicly
lashed out at the Bonus Marchers and at the Legion. The best solution in
Mitchell's mind was the election of a liberal Democrat in 1932, and he
would do all he could to ensure that victory. Mitchell attended the
Democratic Convention in Chicago and had a brief meeting with FDR
where he suggested that Senator Harry Byrd of Virginia be considered for
vice president. As a member of the Virginia delegation, he stated the obvious, that Virginia and the South would look even more favorably upon the
Democrats in 1932.[21] Despite his delegate position with the state of
Virginia, there were still lingering questions about Mitchell's true state
affiliation—was he a Virginian, or a Wisconsinite who sat with the Virginia
delegation?

In July Mitchell stumped in Virginia and Maryland for Roosevelt,
which brought an enthusiastic note from the candidate for his efforts.[22] In
late August and early September Mitchell went to Albany to meet with
FDR and then appeared in Maine and the rest of the New England states,
where he found that the Republicans were making gains in Massachusetts
and Connecticut. Being always bold, Mitchell advised "My Dear Franklin"
that Al Smith should be prevailed upon to tour all of New England to rally
the party faithful and the undecided. From the Northeast Mitchell traveled
on to Wisconsin, spending several weeks there and in Illinois for campaign
stops. He addressed the issue of the wartime bonus payments and urged that
they not be paid now: "If the whole bonus were paid in cash now, the
majority would spend it within a month. The longest time it could keep a
family would not exceed six months," Mitchell told Roosevelt.[23] Disre-
garding the severe privations many veterans faced, Mitchell felt it was the
political thing to do because, as he saw it, "the majority of thinking people
are interested in, first, the reduction of expenses of the Federal and state
governments."[24] This was a politically ambitious Mitchell who preferred to
ignore the plight of the doughboys who had fought the Great War, many of
whom had served in the U.S. Air Service with Mitchell on the Western
Front. His judgment was harsh, but he had expressed the same type of cal-
loused attitude toward Thurman Bane in 1920 and 1921, and it was, if nothing else, at least consistent with his character.

Those working to elect Franklin Roosevelt were well aware that Billy
Mitchell still had many supporters throughout the country. FDR and his
strategists were happy to use Mitchell for their purposes, and Mitchell's
friends made it clear that they wanted him in the government. An editorial
appeared in *The Service News* in September that lumped Mitchell in with
three retired generals and admirals called the "four aces of national

defense" and urged that Mitchell be considered for Roosevelt's "official family."[25] *The Service News* had wide distribution among military personnel, and many of them sent copies to Mitchell. When Franklin Roosevelt won the election, Billy Mitchell had every reason to be pleased with his role in the successful campaign. James A. Farley, one of FDR's trusted advisers, sent Mitchell a personal note of thanks for his work, especially in Virginia and Maryland, but said nothing beyond that.[26] Mitchell had repeated his efforts in the 1928 campaign by speaking and making appearances, and all out of his own pockets, which were not very deep. He contracted several loans while in Milwaukee from the Central Investment Company, which had relative Washburn M. Becker as vice president, and those loans would come due early in 1933.

It is not clear what FDR had in mind when dealing with Mitchell. After his victory Roosevelt wrote to Mitchell, "You gave me so many tantalizing glimpses of a subject on which you are certainly well qualified to speak that I was tempted to set an immediate date for the talk which you suggest. However, I am afraid I shall have to adhere strictly to my own rule not to turn aside from the duties of the Governorship to consider matters of federal moment. After January first we must have that talk."[27] Roosevelt had a mastery of the English language, and his letter was vague enough to offer some hope to Billy Mitchell that his campaigning would give him influence. But, would FDR be willing to have a disruptive influence in his government when there was so much to do? Mitchell certainly believed that he was now a player in FDR's administration. He wrote to FDR before Christmas 1932, "I shall be back home around the first of the year, and shall hold myself ready to come up and see you at any time and bring the data I have about the reorganization of the departments and particularly about our aviation development."[28]

Despite pockets of support for Mitchell's appointment to a governmental post, grave problems with Mitchell's residency would cause some to think twice about endorsing him. Mitchell had never given up his Wisconsin residency, but he had been active in Virginia politics, and, consequently he appeared to be a man with no definite roots. Joseph Davies, a prominent Washington lawyer who had represented Margret Ross Landsdowne during the 1925 *Shenandoah* crash hearings, was a key figure in Democratic politics. Davies was powerful enough that FDR would appoint him ambassador to Moscow in 1936. He warned Mitchell that he was in trouble, even as supporters sent in letters for Mitchell, and that in Virginia Democrats might view a Mitchell appointment as a slight to Senators Harry Byrd and Carter Glass. "You are living in Virginia, and must have the good will of the political leaders in your state or it might prove very embarrassing and possibly fatal before we knew it," Davies wrote.[29]

Mitchell did not do his Virginia homework well and continually relied on letters to FDR. In early January he met briefly with the president-elect in New York City, where he gave him a chart for a department of national defense with an independent air force on an equal footing with the army and the navy. He followed up the meeting with a letter with wide, sweeping proposals for a complete reorganization of all of the departments of government into six agencies, doing away with the Departments of Commerce and Agriculture and the Post Office Department. "Our National Defense Act should be entirely revamped," Mitchell told FDR.[30] Mitchell firmly believed that he had great influence with Roosevelt and often spoke as if the new administration would adopt his proposals. Dr. Joseph S. Ames, Chairman of the National Advisory Committee for Aeronautics, sent, as a courtesy, a copy of proposed changes as to how the committee conducted business, including acting as an adviser to the secretary of commerce.[31] Mitchell's response to Ames was extraordinary: "As far as the Department of Commerce is concerned, when *we* [my emphasis] get through with it there won't be much of it left."[32] Mitchell did add that he did not know exactly what measures Roosevelt would take, but he was certain that "decided changes will be brought about." How Mitchell knew this was something of a mystery because Roosevelt had discussed no real specifics with him. The president-elect listened to Mitchell, read his letters, and glanced at his charts, but certainly promised nothing.

Mitchell's exaggerated estimation of his position was buoyed even more when he was named vice chairman of the Committee for the Reception of the Governors of the States and Distinguished Guests for FDR's March inauguration.[33] Referring to him as General Mitchell, Cary T. Grayson, chairman of FDR's inauguration committee, asked Mitchell to also serve on the general inaugural committee.[34] Billy Mitchell agreed immediately to accept both positions. By the end of March 1933, however, Mitchell found that obtaining an audience with the new president was becoming difficult. He told Arthur Brisbane that he expected very shortly to meet with FDR to discuss the department of national defense.[35] FDR, however, continued to put off any meeting with Billy Mitchell, and by September President Roosevelt himself wrote to Mitchell postponing any meeting for the rest of 1933.[36]

Mitchell went to Capitol Hill in late March to testify for a unified air service and a department of national defense. It was the old Mitchell claiming that, "Battleship building is a racket." With Generals Fechet and

Foulois in the audience, Mitchell told the congressmen, "Airplane carriers are merely a delusion. Airplanes can go out and put them down with a little bomb." As if that were not enough, Mitchell went on to criticize the American Legion lobby as disgraceful in its efforts to get the bonus paid off during hard economic times. "They are going around threatening Congressmen, when in fact they could not carry a single election district," he was reported to have said.[37] When Fechet and Foulois testified they disputed Mitchell's assertions as unsupportable, and the American Legion, once a pillar of Mitchell support, was furious with him. Mitchell's generalizations and statements about "a little bomb" hurt his chances more than he ever realized. Billy Mitchell also had a tendency to lean back in his chair and wave his hand as he spoke, giving the impression that he was dismissing the unlearned. Those who watched him saw an arrogant Mitchell, a know-it-all Mitchell, and that was not lost on the now-majority party Democrats on committees.

By July 1933, Mitchell had begun to realize that he was not the player in the national defense debate he thought he would be. In a letter to FDR, written on Independence Day, he said, "The time has come, I believe, when we should discuss the matter of the organization and handling of American aeronautics. . . . I realize how very busy you have been but I think that the formulation of a sound policy with regard to American aeronautics is important."[38] It was important to Roosevelt as well, but not in the way Mitchell believed it would be. Mitchell's letter to FDR was sent as a matter of form to the Commerce Department, which sent Mitchell a bland, pro forma reply.

While Mitchell was waiting for FDR to grant him an interview, his financial condition became a matter of serious concern. A collection agency in New York, acting on behalf of his London Saville Row tailor, sent a number of letters requesting payment for his hunting clothes and threatening legal action.[39] A realty company to whom Mitchell already owed $13,000 turned him down for an additional $4,000 loan. The Milwaukee company had loaned the amount on Mitchell's Wisconsin property, but could go no further.[40] Mitchell also owed almost $1,000 to a Virginia supplier of feed and grains and was having a difficult time paying that note.[41] He was experiencing a drought publishing his articles, for which he asked about $150 per submission.[42] *Liberty* magazine—an old Mitchell standby—turned down an article, which came as a nasty surprise for Mitchell.[43] He submitted an article on Alaska and Japan and the threat of war to *Field and Stream*, whose editors promptly rejected it as not really the type of article their readers cared to see.[44] *Woman's Home Companion* accepted an interesting article, "The Automobile of the Air," which focused on the autogiro, or early helicopter. In typical Mitchell fashion he claimed that very soon the autogiro would allow its owner to "climb in your front yard, start the propeller, run a few feet, then climb at an angle of fifty degrees and fly to your office building downtown."[45] He cited the 1923 work done by Juan de la Cierva, and then went on to make extravagant claims for the autogiro as the "most versatile and useful means of heavier-than-air transportation."

The picture of Billy Mitchell on New Year's Day 1934 was a sad one indeed; in debt, articles rejected, and no word from FDR about an interview. Mitchell again tried to contact FDR in January 1934, writing, "I would like very much to talk to you about the simplification of our government departments, our national defense, and particularly our aeronautics. . . . Our aeronautics deserves very serious consideration, a decided change in policy and a complete revision. . . . Naturally I have considered the fact that you have been very busy over other matters of great importance up to the present, but I believe the time has come when something must be done about the matters."[46] Mitchell addressed this letter to "My dear Franklin," but Franklin Roosevelt did not personally reply.

In early March Mitchell was back in the headlines when he spoke at a luncheon of the Foreign Policy Association at the Astor Hotel in New York. He claimed that private airlines were unsafe, warning that "the passenger planes they have provided do not have the necessary modern safeguards." He attacked the Curtiss-Wright Corporation as controlling the industry and called them "aviation profiteers."[47] Thomas A. Morgan, president of Curtiss-Wright, was furious and demanded that Mitchell retract his "utterly false" charges about the unsafe condition of passenger planes and that the corporation was in collusion with banks to monopolize the aircraft industry.[48] Morgan could see that Mitchell would not retract his bold statements, and he directed his lawyers to sue Billy Mitchell in court. Mitchell sent a letter to Morgan refusing to take back anything he had said in New York, and he told the press that "I do not know the nature of their charges and it does not worry me. I will defend the suit and we will thresh it out in court."[49] It never came to that, because within two years Mitchell would be dead. But the lawsuit put another burden on the precarious Mitchell finances, a burden Billy Mitchell could not afford. The suit also damaged Mitchell's credibility as a potential Roosevelt appointee.

Mitchell appeared to be uncontrollable when he went before Congress in late March. Franklin Roosevelt had decided to do something about the poor state of airmail in the United States, and he canceled contracts. Secretary of War George Henry Dern called a committee consisting of Charles Lindbergh, Hugh Drum, Benjamin Foulois, and Orville Wright. Lindbergh declined Dern's call, citing his opposition to the military doing

what business should do. FDR brought Mitchell into the picture "as a coun-
selor of the President, a luncheon guest at the White House, and a witness
for the [Senator Hugo] Black Committee."[50] If Roosevelt ever entertained
thoughts of bringing Mitchell into the government, the latter's behavior and
intemperate language put a quick end to them. Mitchell also angered Dern,
and the White House received letters expressing great concern over
Mitchell. Edgar Staley Gorrell, one of the most respected Air Service officers of the Great War and president of Stutz Motor Company, wrote, "With
only the ultimate success of Aeronautics at heart I feel that I owe it to you
to place before you the undesirability of appointing General Mitchell to any
such responsible Governmental aeronautical position. . . . I possess what in
my opinion is positive knowledge as to General Mitchell's unfitness for any
senior governmental Aeronautical position which knowledge I am willing
to place at your disposal. . . . I feel it my duty to respectfully suggest that
you investigate official records on file in Washington in regard to his
career."[51] Roosevelt was in the process of selecting a board to examine aviation, but Mitchell's name would not be on the president's list. Respected
Democratic congressmen like J. J. McSwain of South Carolina and Joe T.
Robinson of Arkansas urged FDR to place Mitchell on the board,[52] but
Roosevelt was too shrewd to have a loose canon on the deck, a figure who
would quickly alienate everyone and grab headlines. In the end, it was
Billy Mitchell himself who killed his chances for greater things.

In April George Dern announced the formation of the Baker Board to
investigate what was needed to improve American aviation. Eleven well-
known public figures were on the board headed by Newton Baker, who had
served as secretary of war during the Great War. Pilots Clarence D.
Chamberlain, James Doolittle, President of Massachusetts Institute of
Technology Karl T. Compton, Edgar Staley Gorell of Stutz Motors,
Research Director of the National Advisory Committee for Aeronautics G.
W. Lewis, and Generals Hugh A. Drum and Benjamin Foulois were among
the board members. Charles A. Lindbergh and Orville Wright were issued
invitations to participate but refused. After the board was complete, its
members went to work with an admonition from Dern that their task was
technical and not political.[53] He was determined that there would be no
Billy Mitchell tirades or headline grabbing during the considerations of the
board. Despite urging from a number of congressmen and senators, Dern,
who had had his fill of Billy Mitchell, refused to let Mitchell get anywhere
near the official board.

In July the Baker Board announced that they would have a report for
FDR before the month was out, and a second commission headed by
Atlanta publisher Clark Howell began an inquiry into a comprehensive pol-

icy for American aviation. Howell was selected by Roosevelt because he came to the inquiry with no specific knowledge about air matters but had a fair, balanced mind, and he pledged that he would consult the Baker Board and other important members of the air community.[54] Mitchell, sitting on the sidelines at Boxwood, was furious over the Baker Board and the Howell Federal Aviation Commission, telling the press that he had been purposely excluded from any discussion over matters in which he had been involved since the Great War. He called both groups a "whitewash." Billy Mitchell was correct in his view that Dern and Roosevelt had excluded him.

The Baker Board rejected the idea of either a unified air service or an independent air force, citing unity of command as essential for the national defense. There were sound proposals for competition of design, building up of the Army Air Corps' equipment and personnel, and emphasis on civil aviation including development of airways, schedules, air navigation, and ground facilities. Much of the well-reasoned military portion was written by James Doolittle, who took care to consider the nation's financial capabilities during a period of recovery.[55] Mitchell disputed the board's recommendations, saying, "This report is about what you would expect of a board packed with army men who know nothing of aviation."[56] When called to testify before the Federal Aviation Commission, Mitchell blamed all of his perceived shortcomings of the Baker Board on Hugh A. Drum and Secretary Dern. By implication Mitchell was blaming Roosevelt for the board, its results, and for rejecting a department of defense.

Mitchell went to work with his typewriter to produce more articles aimed mainly at the Baker Board, but much of his time had to focus on his personal finances, which were in great decline. To Arthur Brisbane of the Hearst papers he wrote in a very strange and self-promoting letter that "since I relinquished direction of our aviation in 1925, a comparatively small group of financial manipulators has gained control of the production of aircraft and engines." Having forgotten that Mason Patrick, not Billy Mitchell, was chief of the Air Service in 1925, Mitchell went on to propound a theory of conspiracy that the small group distributed stock to influential people throughout the country to maintain control of the aircraft industry for their own profit. "The money that maintains them is from 70% to 90% government money, that is they are dependent on the government for their existence." Who were these people? Mitchell cited Elliott Roosevelt, the president's son; Thomas Morgan; Clark Howell; and others.[57]

Mitchell's letter to Brisbane then turned to the lawsuit Thomas Morgan and Curtiss-Wright had filed against him, which was costing him a great deal of money. He had retained three prominent attorneys—Joseph E.

Davies, Seth Richardson—a former assistant attorney general—and Leo P. Harlow, of Alexandria, Virginia. The lawsuit, in Mitchell's views, "was to prejudice the President against me, to discredit me in the eyes of the people, and to take up so much of my time and money in fighting them that my activities in aviation would be necessarily limited." He then asked Brisbane if it "would be feasable [sic] to call on the American people to help me in this matter in a financial way. It occurs to me that it might be possible through the Hearst chain of newspapers to point out some of the facts I have mentioned above . . . and to ask that everyone contribute whatever they feel like, from 25¢ up."[58] This was an extraordinary letter from Mitchell showing the depths of his frustration at being out of the debate over airpower. Isolated and threatened by the Curtiss-Wright lawsuit, Billy Mitchell had basically become irrelevant in the national discussions over air policy. For the first time since 1919 Mitchell, "the stormy petrel of the air," was out of the limelight.

In January 1935, T. V. Rank of the Hearst papers returned a number of articles to Mitchell, which he stated were nothing more than "scathing editorials. They simply denounce the Baker Board rather than review the findings."[59] Mitchell responded to this with a petulant letter that effectively ended Mitchell's relationship with the Hearst chain. In July, Foulton Oursler of *Liberty* magazine, a familiar venue for Mitchell, rejected all of his articles, saying, "The fact is that the aviation song had grown a little tiresome to our readers and we have to give them what they want."[60] To complicate his life even more his sister Ruth had received medical care from a New York physician, had not paid her bills, and Mitchell was asked to assume them. In a notation on the letter Mitchell simply stated that the bill should be referred to his sister Harriet in Milwaukee.[61]

Billy Mitchell still had a few friends on Capitol Hill, and most of those, like Senator Robinson of Arkansas, remembering Mitchell's support for the 1928 Smith-Robinson ticket, knew of his declining financial circumstances. In the late summer of 1935 Robinson introduced a bill that would have authorized President Roosevelt to grant Mitchell the permanent rank of brigadier general and give him his retired pay as a colonel. At the same time Congressman J. J. McSwain of South Carolina, a longtime friend of Mitchell's, announced that he would introduce a similar measure in the House. Since McSwain was the chairman of the House Military Affairs Committee, it was considered that Mitchell had a very good chance of seeing the bill pass. No one was prepared for Dern's vehement opposition to the bill, however. Dern, who would not have been so public had he not discussed it with FDR, stated that in 1926 when Mitchell resigned he stated that the army "owed him nothing," and if that were the case in 1926 it would be so in 1935. On January 25, 1936, less than a month before Mitchell's death, McSwain announced that the committee declined to pursue the matter further. "Billy Mitchell has a lot of friends on the committee," one congressman said, "but there were other phases of the question to be considered besides restoring him to the retired list."[62]

It could not have mattered much to Mitchell because he was in the last fight of his life in a hospital in New York City. He had planned a trip to the city for some time, but when he arrived there he felt ill, and on February 9, 1936, he was admitted to the Doctors' Hospital with a severe case of influenza. Mitchell had been in good physical health, and when interviewed in Washington in the fall of 1935 he seemed to be in fine shape, full of fight, the old Billy Mitchell. But influenza, which had killed so many doughboys of the Great War in 1918 and 1919, was still in the 1930s a serious and little understood disease with no known effective medical interventions. On February 11 and 12, his doctor, Samuel W. Lambert Jr., reported that Mitchell was "a little better," but his words were guarded. Betty Mitchell gave no interviews to the press, but it was clear to all that Mitchell was probably not going to win this battle. Sinking slowly into delirium from the effects of the virus, Mitchell passed a fretful night on February 18, and the next morning, February 19, 1936, Billy Mitchell died quietly without ever regaining consciousness. It was an irony that the "D'Artagnan of the Air,"[63] a man who had occupied headlines for fifteen years, the "Intrepid Airman" who sunk the *Ostfriesland,* would leave this earth without one last call for his dream of a uniform air service or department of national defense. Instead of one great air battle to settle things, Billy Mitchell died from a disease that had killed so many infantrymen in the war.

President Franklin Roosevelt observed Mitchell's passing and said the United States government would send a wreath to St. Paul's Episcopal Church in Milwaukee. Secretary of War Dern said nothing publicly, nor were there comments from James E. Fechet or Benjamin Foulois. The chief of staff of the army, General Malin Craig—one of Mitchell's foes from the war—stated only that Mitchell's service in the Great War would be remembered. Like many of his colleagues in uniform Hugh A. Drum tastefully said nothing. That was not the view of many in Congress who were still his friends. In the Senate in March 1936, less than a month after Mitchell's death, a bill was passed awarding him the Congressional Medal of Honor for his wartime services.[64] In April the House passed a similar measure, but the whole idea of the Medal of Honor for Billy Mitchell died because of strong opposition from Dern. The War Department pointed out that Mitchell was awarded the Distinguished Service Cross for gallantry during the war, and documentation for a Medal of Honor would be hard to assemble.

On March 26, 1936, Elizabeth Miller Mitchell went to court to settle Billy Mitchell's will and estate. The court in Warrenton, Virginia, noted that he had owned no Virginia property, as Boxwood was the property of Betty Mitchell. His personal estate amounted to $5,765 in accounts and cash; his personal collection of firearms and hunting trophies were valued at $15,000, and they went to Elizabeth. No mention was made of his three children by his first marriage to Caroline Stoddard Mitchell.[65]

NOTES

1. Gwen Dobson (ed.), *Middleburg and Nearby* (Private Printing by the Fauquier County Historical Society, 1986), 263.

2. Eugene M. Scheel, *The History of Middleburg and Vicinity* (Warrenton, VA: Fauquier County Historical Society, 1988), 125–126.

3. *New York Times,* 2 November 1928, 4.

4. Jenkins to Mitchell, 23 May 1929, in the William Mitchell Papers, Library of Congress, Washington, DC, Carton 14. (Hereafter cited as the Mitchell Papers.)

5. William Mitchell, "America in the Air," *National Geographic* 39, 4 (March 1929), 339–351.

6. Mitchell's investing can be found in notes in the Mitchell Papers, Cartons 12 and 13.

7. Ziegler to Mitchell, 8 December 1929, Mitchell Papers, Carton 14.

8. Mitchell to Patton, 9 December 1929, Mitchell Papers, Carton 14.

9. Patton to Mitchell, 11 December 1929, Mitchell Papers, Carton 14.

10. Mitchell to Ziegler, 16 December 1929, Mitchell Papers, Carton 14.

11. For examples, see Henry A. Arnold, "The Cavalry-Air Corps Team," *The Cavalry Journal* (January 1928); and Edward M. Fickett, "A Study of the Relationship Between Cavalry and the Air Service in Reconnaissance," *The Cavalry Journal* (October 1923).

12. William Mitchell, "A Glance at World Aeronautics," *Saturday Evening Post* 202 (19 April 1930), 6–7, 66.

13. William Mitchell, "The Sporting Side of Aviation," *Saturday Evening Post* 202 (30 April 1930), 37–41, 157.

14. Mitchell to Michael Cudahy, 17 July 1931, Mitchell Papers, Carton 16.

15. Veterans Administration Form, Claim 1706746, 2 December 1931, Mitchell Papers, Carton 50; Pension Granted, 10 May 1932, Mitchell Papers, Carton 50.

16. Mitchell to Brisbane, 1 December 1931, Mitchell Papers, Carton 50.

17. Rank to Mitchell, 8 August 1932, Mitchell Papers, Carton 50.

18. Mitchell to Rank, 20 September 1932, Mitchell Papers, Carton 50.

19. Ruth Mitchell, *My Brother Bill: The Life of General "Billy" Mitchell* (New York: Harcourt, Brace, and Co., 1953), 340–341. Though this book must be used with caution there is ample evidence that FDR did hint at a position.

20. Scheel, *Middleburg,* 129–130.

21. Mitchell to Henry Breckenridge, 12 July 1932, Mitchell Papers, Carton 16.

22. Roosevelt to Mitchell, 1 August 1932, Mitchell Papers, Carton 16.

23. Mitchell to Roosevelt, 29 September 1932, Mitchell Papers, Carton 16.

24. Ibid.

25. Editorial, *The Service News* (16 September 1932), Mitchell Papers, Carton 16.

26. Farley to Mitchell, 17 November 1932, Mitchell Papers, Carton 16.

27. Roosevelt to Mitchell, 19 November 1932, Mitchell Papers, Carton 16.

28. Mitchell to Roosevelt, 21 December 1932, Mitchell Papers, Carton 16.

29. Davies to Mitchell, 27 December 1932, Mitchell Papers, Carton 16.

30. Mitchell to Roosevelt, 17 January 1933, Mitchell Papers, Carton 16.

31. Ames to Mitchell, 12 January 1933, Mitchell Papers, Carton 17.

32. Mitchell to Ames, 17 March 1933, Mitchell Papers, Carton 17.

33. Ray Baker to Mitchell, 19 January 1933, Mitchell Papers, Carton 18.

34. Cary T. Grayson to Mitchell, 18 February 1933, Mitchell Papers, Carton 18.

35. Mitchell to Brisbane, 30 March 1933, Mitchell Papers, Carton 17.

36. Roosevelt to Mitchell, 28 September 1933, Mitchell Papers, Carton 16.

37. *New York Times,* 31 March 1933, 27.

38. Mitchell to Roosevelt, 4 July 1933, Mitchell Papers, Carton 16.

39. McKillop, Walker, and Co. to Mitchell, 6 May 1933, Mitchell Papers, Carton 18.

40. Mitchell Mackie to Mitchell, 29 August 1933, Mitchell Papers, Carton 18.

41. R. S. Cochran to Mitchell, 8 June 1933, Mitchell Papers, Carton 17.

42. B. W. Schreiber to Mitchell, 18 July 1933, Mitchell Papers, Carton 16.

43. Elliot Balestier, Associate Editor, to Mitchell, 20 April 1933, Mitchell Papers, Carton 16

44. R. P. Holland, Editor, to Mitchell, 4 August 1933, Mitchell Papers, Carton 16.

45. William Mitchell, "The Automobile of the Air," *Woman's Home Companion* 59 (May 1932), 18–19, 126–127.

46. Mitchell to Roosevelt, 5 January 1934, Mitchell Papers, Carton 19.

47. *New York Times,* 4 March 1934, 3.

48. Ibid., 6 March 1934, 4.

49. Ibid., 17 February 1934, 7.

50. *Literary Digest,* 24 March 1934, 7.

51. Gorrell to Roosevelt, 13 March 1934, Mitchell Papers, Carton 19.

52. McSwain to Roosevelt, 14 March 1934; and Robinson to Roosevelt, 27 June 1934, Mitchell Papers, Carton 19

53. *New York Times,* 18 April 1934, 5.

54. Ibid., 11 July 1934, 17.

55. Ibid., 23 July 1934, 6.

56. Ibid., 24 July 1934, 6.

57. Ibid., 3 October 1934, 4.

58. Mitchell to Brisbane, 4 September 1934, Mitchell Papers, Carton 19.

59. T. V. Rank to Mitchell, 20 January 1935, Mitchell Papers, Carton 19.

60. Oursler to Mitchell, 31 July 1935, Mitchell Papers, Carton 19.

61. Dr. W. S. Bainbridge to Mitchell, 21 January 1935 (Mitchell's note on letter dated 4 February 1935), Mitchell Papers, Carton 19.

62. *New York Times,* 29 January 1936, 3.

63. Helen S. Waterhouse, "D'Artagnan of the Air," *Christian Science Monitor* (9 October 1935), 4–5.

64. *New York Times,* 13 March 1936, 21 and 25 April 1936, 21.

65. Ibid., 26 March 1936, 3.

FOURTEEN

MITCHELL REVIVED

E VEN IN DEATH BILLY MITCHELL REMAINED CONTROVER-
sial. When his body was taken off the train from New York that cold
day in February 1936, a different storm was brewing in Washington. On
Thursday, February 20, noted radio news broadcaster and commentator
Boake Carter blasted Chief of Staff Malin Craig and the army General
Staff, saying that they personally had refused to bury Mitchell in Arlington
National Cemetery among the nation's honored war dead. Capitol Hill and
the White House were deluged with telegrams and phone calls protesting
this alleged ill-treatment of a Great War veteran. Roosevelt and the White
House staff were furious over the broadcast, and Malin Craig issued a statement to the press confirming that "there is no reason why General Mitchell
should not have been buried in Arlington national cemetery had he or his
family had expressed such a desire, and the war department would have
extended to him the same military honors it gives to any veteran of honorable service."[1] Craig then sent a telegram to Carter stating, "The war
department requests that you correct the impression you created, giving
publicity to the true facts over the air in the same manner as the original
error was broadcast."[2] This a very embarrassed Carter did that very day, but
the damage was done to the reputation of the army, and the idea that in
retaliation for his outspoken positions in 1924 and the trial in 1925 the
army denied him a veteran's honors remained in Mitchell lore.

On Saturday, February 22, 1936, Billy Mitchell was laid to rest in
Forest Home Cemetery in Milwaukee, near his grandfather and father. His
flag-draped coffin was open, and old friends passed by to pay their last
respects. Betty Mitchell received telegrams from Craig, Pershing, Foulois,
Fechet, and others in the army. Though President Roosevelt had announced
that the nation would send a wreath to honor Mitchell's services in the

275

Great War, following Carter's radio broadcast the president sent General A. W. Robbins to Milwaukee as his personal representative. Major General Frank McCoy, commander of the VI Corps headquartered in Chicago, announced that he would attend the funeral and would be honored to act as one of the pallbearers at the funeral. His aide, Colonel George C. Marshall, attended the service with his commanding officer and also acted as a pall-bearer.[3] Both McCoy and Marshall had known Mitchell very well from their Washington service after the Great War because they all shared a love of horses, and both officers had been riding companions of Mitchell in better days.

Realizing the storm caused by the Carter broadcast, Betty Mitchell issued a statement quoting Mitchell's own desires about burial: "Although I should like to be with the pilots and my comrades in Arlington, I feel that it is better for me to go back to Wisconsin, the home of my family." Four members of the House Military Affairs Committee tried to fly from Washington to Milwaukee, but their airplane was forced down near Winchester, West Virginia. After repairs were made to the plane, a shaken Chairman J. J. McSwain of South Carolina, John M. Costello of California, Andrew Edmiston of California, and Matthew J. Merritt of New York continued their flight and arrived at St. Paul's Episcopal Church in time for the service, which included the hymns "Lead Kindly Light" and "The Battle Hymn of the Republic." Mitchell's body was then taken to the cemetery for a rather unmilitary graveside service with only an American Legion firing squad to fire the last volley and a bugler to sound "Taps."[4] None of Mitchell's children by his first marriage attended the funeral.

Elizabeth Trumbull Miller Mitchell returned to Boxwood and over a period of a few months settled her late husband's modest estate. After a few years she married Thomas B. Byrd of Boyce, Virginia, the brother of the famed Virginia Senator Harry Byrd and Admiral Richard Byrd, the man who Mitchell fought so bitterly in 1926 over the presidency of the National Aeronautical Association.[5] In 1943 Boxwood was sold to a Mr. and Mrs. Christopher Greer, who in turn sold it in the 1970s. The present owners of Boxwood in 1975 petitioned to have the 1826 farm placed on the National Historical Register, which it was. It came to be known locally as the "General Billy Mitchell House."[6] The house and grounds stand as they were when Billy and Betty Mitchell made their home there from 1926 to 1936.

After his death Mitchell's name faded from public view until war clouds began to gather in Europe and in Asia. In October 1940 the *Chicago Tribune* carried an article about an experimental parachute unit being formed at Fort Benning, Georgia, from volunteers of the 29th Infantry

Regiment. The first call for volunteers to undergo parachute training was made on June 29, 1940, and was a response to spectacular successes that Nazi German paratroopers had experienced in the recent fighting in Europe.[7] As the American soldiers underwent exhausting training to prove the validity of the Parachute Test Platoon, the *Tribune* called for the projected 500-soldier battalion to be called the "Mitchell Battalion." During the Great War and, again in 1928, he had surfaced the idea of parachuting an American unit drawn from the 1st Infantry Division behind German lines. He argued during the war and thereafter that the unit could be supplied by air, supported by attack aircraft, and act in concert with tanks and infantry attacking from the front.[8]

While it is true that Mitchell suggested this to Pershing in October 1918, it was a bit of a stretch, as the *Tribune* editorialized, to say that Mitchell was the father of the paratroops and that he had looked into the future and foreseen the Nazi blitzkrieg. First of all, the method for sending large numbers of parachute infantry into battle quickly did not exist in 1918. Aircraft did not carry parachutes although balloons did, and there was no precedent or doctrine for an airdrop of battalion-size units. The means to resupply a large combat unit from the air was also a dream of the future during the Great War. When aircraft went to assist the 77th Division's famed Lost Battalion during the Argonne fight, they had trouble finding them and could do little more than drop messages to the embattled group of doughboys. It was very easy to look back on what Mitchell proposed and then make him the prophet of the airborne operations in 1940. The reality of 1918, or even 1928, was that no general would have committed troops to what would most probably have been a suicide mission and most likely could not have been carried out anyway. By 1940 the changes in technology in aircraft and in parachutes had made the Parachute Test Platoon a realistic experiment that Chief of Staff George C. Marshall called for in light of events in Europe a few months before.

The invasion of Poland in 1939 and the spectacular German victories in France in the spring of 1940 showed that the Germans had developed ground and air combat forces vastly superior to those of the former great powers. When the battles in France broke out in May and June 1940, the concentration of airpower and the effective use of the Stuka dive bomber in combination with highly mobile ground forces made victory over Britain and France look frighteningly easy. The reaction of American military planners, especially George C. Marshall, was as one might expect from a group of professional officers. They became concerned as they read the lessons learned from German blitzkrieg methods, and they began to make major alterations in the composition of American ground and air forces.[9] Staggered by the swift collapse of France and the ensuing Battle of Britain, the American public craved an explanation of what had led to these troublesome events in Europe. In 1940 and 1941 a large number of popular writers and journalists began to "vindicate" Billy Mitchell. Almost every writer reviewed Mitchell's calls for a unified air service and a secretariat of defense to direct American military efforts. His articles on America's vulnerability were dusted off and related to the terrible air bombardment that Britain was sustaining as they wrote.[10] Even before Pearl Harbor, Billy Mitchell was taking on the mantle of the prophet, the great war thinker whose warnings had fallen on deaf ears. In many ways they were right, because Mitchell was the officer who had massed over fourteen hundred combat aircraft during the St. Mihiel operation. His 1919 paper on the tactical use of airpower stressed joint air-ground operations. Very few of the popular writers of 1940–1941 understood the nuances of Mitchell's Great War achievements and preferred to concentrate on his postwar testimony and his writings.

As another great war swept over Europe it was impossible to keep the Mitchell name out of the newspapers, and Ruth Mitchell continued the "old family tradition." When the Wehrmacht moved into Yugoslavia in early spring 1940, Ruth Mitchell, who was in Yugoslavia ostensibly to write a travel guide book, was arrested and put in a German concentration camp because of her work with anti-Nazi Serbian patriots. Though she denied the charges, subsequent events would show that she did indeed assist democratic forces in subjugated Yugoslavia. By the winter of 1941 the State Department and the White House were deluged by letters and telephone calls to assist in obtaining Ruth Mitchell's release.[11] Through the good offices of the Swedish ambassador in Berlin, Ruth Mitchell was finally released; she returned to the United States and then went on a speaking tour denouncing Nazi atrocities in the Balkans and supporting Serbian anti-Nazis.

It was the Japanese attack on Pearl Harbor, Hawaii, on December 7, 1941, that brought even more attention to Billy Mitchell and his ideas. The United States was embarrassed at the devastation wrought by Japanese naval aviation in such a short time in Hawaii and their rapid, decisive destruction of Philippine air facilities. Almost immediately journalists and military writers began to point to Mitchell's sinking of the *Ostfriesland* in 1921 as an example of his warnings to the United States. Reports of his tour of the Pacific in 1923 and 1924 were dusted off, and it was clear that Mitchell had been quite accurate in describing what could happen if a hostile power were to attack the Hawaiian Islands and the Philippines. His report concerning the sorry state of air defenses on the Philippine Islands well mirrored what the United States now knew of preparations in that area. The movements of the Japanese to take the islands around Alaska promoted the editorial writers of the Columbia, South Carolina, *State* to write, "The late General Billy Mitchell, who in recent years has hung up a pretty good batting average as a prophet, said, 'He who holds Alaska holds the world.'"[12] Mitchell had been consistent over the years in warning that in a war in the Pacific, Alaska and the Alaskan Islands would pay a major strategic role for the United States.

The worst blows fell on the navy at Pearl Harbor, with so many stationary battleships sunk in such a short time. Mitchell's *Ostfriesland* triumph and his subsequent fight with the navy made good sources for journalists trying to make sense out of the new type of war Americans were now called upon to fight. With so many of the Pacific Fleet's battleships a smoking ruin, it was clear that somebody had to have misjudged what the nation's defensive capabilities actually were. There was a storm of protest and an avalanche of journalistic criticism when it became known that in 1940 the Navy Department and Congress had authorized the building of five new Montana Class super-battleships, named Montana, Ohio, Maine, New Hampshire, and Louisiana. When in the aftermath of Pearl Harbor President Roosevelt and Congress put the new battleships on hold, editorial writers trooped out Billy Mitchell. The *New York Journal American* in the spring of 1942 editorialized that it was Mitchell who had warned long before that "the next war would be fought in the air and nations best prepared in military aviation would win."[13] The naval appropriations bill reported out of Congress set aside the enormous sum of $8.5 billion, but none of it was earmarked for new battleship construction. Citing the tremendous outlays in steel and other wartime-short materials, President Roosevelt finally put an end to the building of super-battleships.

Mitchell had been right as far as the battleship was concerned, but his fight with the navy went beyond just hammering away at the great ships. Much of Mitchell's fight was a power grab, trying to gain more and more control over national defenses for the Air Service. His fight with the navy from 1919 on, while correct in many respects, was a fight for dominance, which tied in with a familiar Mitchell refrain—a cabinet-level secretary of defense, with air being coequal with army and navy. Naval aviation thinkers like Admirals Sims, Fullam, and Moffett could have contributed greatly to Mitchell's arguments against a battleship navy. Mitchell did maintain a cordial correspondence with Fullam, but he could never fully exploit the relationship to mount a united front for airpower. It was beyond Billy Mitchell to share the limelight with anyone, especially one wearing navy blue. Mitchell tried hard to gain control of the lighter-than-air ships, despite a mandate by Congress giving control of the dirigibles to the navy. His reasons for the national defense were sound: stop the enemy's invasion fleets in the waters miles from American shores. His methods obscured his message.

Mitchell never really comprehended the development of the aircraft carrier. In 1921 Mitchell argued correctly that an aircraft carrier could be most effective in the destruction of an enemy fleet far from American shores. But in testimony he revealed that his own thinking about this stand-off defense was that the carrier made a fine vehicle for his proposed independent airforce. The implications were obvious. A new unified air service would operate seagoing vessels when carrying aircraft that belonged to this proposed unified air arm. When testifying before Congress in February 1921, Mitchell had requested that the captured giant German liner *Leviathan,* which had transported so many doughboys to and from France during the war, be converted into a carrier with a 1,000-foot deck. When a shocked congressman pointed out that such a ship could not possibly pass through the Panama Canal, Mitchell responded that the ship could be "constructed with an accordion like hull that would adjust to fit the canal's locks." A prominent naval aviation historian has pointed out that Mitchell's ideas were absurd and showed that he frankly knew nothing about how a modern ship was constructed.[14] But it was vintage Mitchell to speak in glittering generalities and fancies that grabbed headlines.

Mitchell had been in Washington long enough to understand the dynamics of interservice competition, especially for funding. He first put on army blue in 1898 and was no novice when he went to the General Staff before the Great War. His all-out rampage against the navy after his return to the capital in 1919 was contrary to everything he had learned. The unwarranted outbursts and statements after the *Shenandoah* crash in 1925 were symptomatic of what Billy Mitchell had become. His constant clashes with the navy since 1919 are best seen in the light of Mitchell and his psychology, his own view of himself. Had one been privy to the letters between Mitchell and his mother in 1919 and 1920, a good eye would have seen danger signs for the years to come. He saw himself as the only officer who could bring about an independent air service and a secretary of defense with parity among army, air, and navy. As a career officer Mitchell's conduct toward Major General Charles Menoher was not in keeping with the discipline of the army. It did not matter that Menoher was a ground-oriented, nonflying artillery officer. What did matter was that Menoher, personally selected by Pershing, was chief of the Air Service and had two stars on his epaulets, not the temporary one star that Mitchell wore.

Regardless of what cause Mitchell claimed to espouse, he was not a good soldier willing to work within the system. In fact, he undermined Menoher whenever possible and made his own office into a thorn in the side of the War Department. The great failure to impose military discipline on Mitchell was John J. Pershing's. During the Great War Black Jack Pershing never hesitated to relieve a general he thought to have failed in his duty, but he could be unfalteringly loyal to one who served him well, and Mitchell had certainly done that in the St. Mihiel and Meuse-Argonne fighting. Pershing had no better fighting Air Service officer than Billy Mitchell. Only when the situation became intolerable was Mitchell ordered to Walter Reed, and later to Fort Sam Houston, Texas. Senator James Wadsworth of New York came to believe that Mitchell was a "spoiled brat" and that was what he was allowed to become.

Mitchell had done well on his own as a lieutenant and as a captain of the Signal Corps, and he had worked fairly well with others when, as a major, he served on the General Staff. Once he had to deal with equals in a wartime situation where there were raging levels of testosterone, Mitchell found it difficult to fit in. He was constantly disappointed that Pershing never made him the AEF's chief of the Air Service. Billy Mitchell in that position could very well have been a disaster, given his fiery personality and ego. He was always best suited to be a warfighter, not an overall administrator and coordinator. He did not get along well with Generals Kenly or Foulois, and only Mason Patrick had the ability to focus Mitchell's attention on the tasks at hand. He clashed with Hugh A. Drum, Dennis Nolan, and Malin Craig and finally won the enmity of Hunter Liggett. Mitchell's attempts to copy Hugh Trenchard's independent air arm gave him a feeling that he was somehow not exactly a part of Pershing's AEF team. The war did not change Billy Mitchell so much as it gave Mitchell's darker side—his ego, his imagined competition with a long-dead father, his boiling-hot ambition—a chance to surface, and once it did it could not be brought under control except for short periods. He was not home long from France when the change in him became very noticeable to Caroline Stoddard Mitchell, and they started on the downward slope toward divorce.

The centerpiece of the Mitchell saga was, of course, his court-martial during the cold waning months of 1925. In reality the trial and conviction were only one part of the whole story, and the verdict and sentence were a forgone conclusion. The real crisis of 1925 came earlier, with his transfer to Fort Sam Houston, Texas, and with his reversion to his actual rank of colonel. The transfer was the army's last attempt to bring Mitchell into line and end his "lawless" behavior. Mitchell loved headlines and the glare of camera lights, and the trial gave to him an opportunity to restate well-worn themes and to be the center of the controversy. His own conduct during the trial—not paying close attention to the work of the defense, taking afternoon rides, and the like—indicated that he knew what the outcome of the trial would be. Once he challenged that paragon of military virtue, Charles Pelot Summerall, Mitchell's military career was over. Every decade of the twentieth century produced "trials of the century," and the Mitchell court-martial was no less a media circus than the Leopold and Loeb or the John Scopes "monkey" trials. The court-martial produced nothing new; Mitchell was covering old ground, ground he had covered in committee testimony or in his own writings. For Billy Mitchell and his ego it was the trial, not the verdict, the testimony, the lawyer-wrangling in the court, that was important. Did Mitchell really believe that his court-martial would produce a groundswell for a unified air force or a cabinet-level secretary of defense? Probably not, but it did set Mitchell up, or so he thought, to be considered for a government post later on. The great historical problem with the court-martial of Billy Mitchell is basically that those who have not seen the whole Billy Mitchell make it the defining moment of his crusade for air-power. It was not that at all.

Mitchell had ingrained in him a sense of frustration with his father's aloofness, the declining family fortunes after 1893, and finally Senator John Lendrum Mitchell's opposition to the Spanish-American War, the vehicle that allowed Mitchell to act on his own and seek his own level of recognition. Willie Mitchell had been raised at Meadowmere and had developed a deep love of horses, dogs, hunting, and the outdoor life. But Billy Mitchell carried his youthful loves to extravagant heights as a man. Perhaps he recalled the days at Racine College when, with precarious family finances in the background, he had to plead continually for some sort of small allowance for spending money? At any rate, the Mitchell persona was always one of wealth and status, the "Mitchell Stables," the fine hunting clothes from London's Saville Row tailors. Few would have ever suspected that Billy Mitchell lived on the edge of financial disaster, relying on loans and gifts from his mother, the generosity (to a point) of banks, and the purse of his second wife, the lovely Elizabeth Trumbull Miller Mitchell. Boxwood, nestled in the beautiful rolling hills of northern Virginia, was really Betty Mitchell's, but Billy enjoyed to the fullest playing the role of the horse-raising, fox-hunting gentleman.

Mitchell's main contribution to the fortunes of the family after 1925 was in the form of his writing, especially the pay for articles for magazines. Mitchell was a less than average public speaker, but he was masterful with the written word. It is well to note, however, that Mitchell's articles were geared toward a popular audience; very few of them appeared in professional journals. Frankly, Mitchell always needed money, and the widely circulated magazines were a good source of that money. After 1930 Mitchell's acceptance rate dropped considerably because he was rehashing old ideas. His tirades after the Baker Board recommendations were made public alienated many of his old supporters in the magazine business, and Billy Mitchell faded from public view. His arguments for a unified air service and a cabinet-level secretary of defense faded away, only to be revived during World War II. In glib, general terms he wrote about the possibility of air attacks on American cities by aircraft or aerial torpedoes. Mason Patrick had alluded to such possibilities as well in the mid-1920s, and the idea of bombs hitting the United States was not especially original with Mitchell. His view of the possible was enhanced by his visit to the Caproni factories in Italy and with talks with Italian air officials.

It would take World War II, the Cold War, and the atomic age to make possible a radical change in the defensive structure of the United States. In the 1920s the political and diplomatic conditions were certainly not critical, and technology had not reached the point where the air was a major contributor to victory in a massive war. Mitchell's ideas were sound, but they were reasonable only after the end of the Second World War. Was he, then, the "founder of the United States Air Force," as some claimed? He was certainly one who saw the potential of a realignment of the armed services and the cabinet, but it fell to others to actually formulate the plans and put them into practice. The role of the newly created Air Force was certainly much different from anything Mitchell could have conceived of.

What then did Billy Mitchell contribute? Though not practical while Mitchell was alive, the concept of an independent air force and a cabinet-level secretary of defense certainly made sense after World War II, as the United States faced global responsibilities as one of the great superpowers and leader of the West in the Cold War. Mitchell's influence on generals like Hap Arnold, Carl Spaatz, and Ira Eaker was tremendous, and many of his ideas came to fruition through them. They were captains when Mitchell was found guilty in 1925, and they were more practical soldiers. Their claim to be heard came not through the press or articles through journals, but in being leaders of the U.S. Army Air Forces in World War II. Carl Spaatz, who had provided men to prepare stables for Mitchell's horses and who had forcefully testified for Mitchell at his trial, became the first chief of staff of the U.S. Air Force. He and his cohorts made Mitchell's ideas work within the frameworks established by Mason Patrick, Benjamin Foulois, and James Fetchet. Mitchell had preached the combat doctrine of mass and achieved it at St. Mihiel; the bombers that flew from England, Italy, and North Africa in World War II also achieved mass and air supremacy, but at a horrendous cost in planes and crews. Mitchell had his disciples in the generals of World War II, but they could work within a system. That, Mitchell could not have done.

There was always a political dimension to what Billy Mitchell did. In the movie *The Court Martial of Billy Mitchell,* Gary Cooper, playing Mitchell, repeatedly stated that he did not want politics brought into what was basically an army problem. Nothing could be further from the truth. Mitchell was born into a political family, and he understood the potential of political connections even though he often did not use them wisely to his advantage. He was never very far from Wisconsin, and later Virginia, Democratic Party politics. The problem for Billy Mitchell was a very human one—he overestimated his role as a political player. He had connections on Capitol Hill with men like James Wadsworth of New York, who later came to view Mitchell as something less than the heroic jouster against dragons, to J. J. McSwain of South Carolina, Hugo Black of Alabama, and Harry Byrd of Virginia. Mitchell saw himself as a figure of national importance, and he never developed a political power base either in Virginia or Wisconsin. There was the possibility of a run for the House or the Senate from Wisconsin, but he allowed that chance to slip away. He was a tireless worker for Al Smith and for Franklin Delano Roosevelt, but it was clear that Mitchell always expected a position in a Smith or a Roosevelt administration. His sister Ruth Mitchell in her 1953 book stated that Mitchell fully expected a post in the New Deal, and personal letters by Mitchell himself bear that out. Al Smith was a forlorn hope, but FDR was something else again. Here it appears that the Democrats welcomed Mitchell's considerable, personally financed efforts, but as Roosevelt came to grips with the nation's massive problems he did not want a loose canon rolling about the decks of the ship of state.

Mitchell had more drawing power as a campaigner during the Al Smith campaign than he did during the first Roosevelt run for the White House. As one of Mitchell's publishing friends tried to tell him, the Depression and the critical condition of the nation overshadowed the air question. As a gesture of thanks, the Democratic Party included Billy Mitchell in its various inaugural celebrations. It certainly appears that Mitchell misread this, and it was equally certain, as it became apparent that FDR had less and less time to read Mitchell's charts and to listen to his positions on national defense, that Mitchell became embittered. The final blow came with Secretary of War George Dern's vehement opposition to any alteration in Mitchell's status as an officer. But much of what happened to Billy Mitchell was his own doing. He believed that he would shape a unified air service and create a cabinet-level secretary of defense, and he firmly held on to the idea that he would be the man to lead such a major transition.

Mitchell was certainly an angry man, but his life was not unpleasant by any stretch of the imagination. Life at Boxwood was interesting, and Billy and Betty entertained many famous people who had an interest in horses and hunting. While very few came to Middleburg to discuss air issues or politics—in fact, many avoided such discussions with Mitchell—the Mitchells were known as charming hosts. Billy Mitchell became involved in local affairs and local history, manifesting an interest in the American Civil War in that area of Virginia. Throughout the 1930s he and Betty traveled often to New York and to New England as well as to Milwaukee. In January 1934, Billy and Betty Mitchell attended the thirtieth annual dinner of The Explorers Club, an organization Mitchell had belonged to for many years. The gathering was held in the famed Astor Hotel in New York, and the menu was indeed superb, with oysters, filet of sole, filet of beef, broiled chicken, fine wines, and excellent brandies. Though not seated on the dais, Mitchell was placed at table six, close to the speakers, a recognition of his status.[15] Mitchell continued to maintain membership in various Northern Virginia and Washington, D.C., hunting and riding clubs. He was also an elected official of the Washington Yacht Club.

His visits to Milwaukee were filled with nostalgia. His adoring sister Harriet, now Harriet Young, had her own daughter named Harriet, who was quite a good rider, and Meadowmere once again boasted fine thoroughbred horses in its pastures. The irrepressible Ruth was now Ruth van Breda, with a son, John Lendrum van Breda. Like his famous uncle the young John Lendrum was a flying enthusiast, and he would be killed flying for the RAF against the Nazis in North Africa in 1941. His mother was in a German concentration camp when her son fell in air battle.[16]

What was missing from the Mitchell picture was contact with the three children from his marriage to Caroline Stoddard Mitchell. There is no evidence that Mitchell had any association with his three oldest children, despite the fact that Caroline and the children had settled in Washington, D.C., and that she remained active in the Chevy Chase Club, the Riding and Hunt Club of Washington, and the National Country Club. In the early 1930s she met Franklin F. Korell, a University of Oregon and Yale Law School graduate who was an assistant to the general counsel of the Treasury Department. On November 9, 1932, they were married, and Caroline settled into being a grandmother with four grandchildren.[17] When World War II broke out William and Caroline's son, John Lendrum Mitchell II, became a private in the artillery and was sent to Cornell University's ROTC unit for training. He had tried to enlist in the Army Air Force, but eye problems kept him from that branch of the service.[18] He received his commission as a second lieutenant and was sent to Pine Camp, New York, for further training. While there he contracted a severe illness and died in late October

1942.[19] There would be no Mitchell—no Stormy Petrel of the Air—flying for the United States Army Air Force in World War II.

It was, oddly, the Second World War that made Mitchell a legend. In searching for sense among the smoking ruins at Pearl Harbor or in the Philippines, Mitchell was revived as a prophet, a spotless self-sacrificing hero who placed his army career on the altar of public service. The focus was on Mitchell, his trial, and his constant demands for a unified air force and a reordering of the American defense establishment. The warts, and there were many, were ignored. Writers tended to see Mitchell as they wanted to and made out of him the knight of the air, which despite his many accomplishments he was not. In 1942 Simon and Schuster published Alexander P. De Seversky's *Victory Through Air Power*, which the Book-of-the-Month Club immediately decided to offer to its readers. The club's president, Harry Scherman, and the entire editorial board wrote a special preface to the club edition extolling this book as important for layreaders. De Seversky, who had worked with Mitchell in the 1920s, surveyed world airpower and championed an independent air force as the key to victory. His full-page dedication to Billy Mitchell stated in part, "He [Mitchell] has emerged not merely as the clear sighted and farsighted apostle of true air power, but as the human symbol of America's air age. . . . His prophetic words and thoughts today, in this hour of supreme test, point clearly and optimistically to the road of VICTORY THROUGH AIR POWER."

The Allies' great victory over Germany and Japan did not come through the application of only one form of combat power. All services united and shed vast rivers of blood to bring about the defeat of German totalitarianism and Japanese expansionism. Unity of effort by the service men and women, the difficulties of maintaining a wartime allowance, first-rate leadership by generals who had to learn the art of war step by step, and the willingness of a homefront to accept rationing, shortages, and the pain of "the telegram" made victory possible. But World War II revived the memory of Billy Mitchell, and since that time he has remained a figure of great importance in the development of airpower.

In 1942 Congress voted to restore Mitchell's rank of major general, dated to the day of his death in 1936. Mitchell's old friend J. J. McSwain of South Carolina also proclaimed that he would sponsor a bill to award Billy Mitchell a posthumous Medal of Honor. Needless to say, soldiers like General George C. Marshall were horrified over awarding a medal that signified gallantry above and beyond the call of duty while facing an armed enemy. That Mitchell had been brave there could be no doubt, and he had been awarded the Distinguished Service Cross to prove it. The issue buzzed about the Congress until 1943, and in 1946 a special medal for Billy Mitchell was approved.[20]

The United States Air Force, created in 1947, needed to establish itself as an equal partner in the defense of America. The Cold War, with nuclear holocaust ever lurking in the background, made it imperative that the USAF begin to function at as high a level of efficiency as possible. To build any military unit there is a vital need for morale and esprit de corps, and what better way to instill this than to show those wearing Air Force blue that they did indeed have a glorious history? So many of its leaders, Carl Spaatz, Curtis Lemay, and others, had come from the old Army Air Corps and the Army Air Force. Billy Mitchell was different. He became one of the shining stars of the new USAF, and for good reason—he had been the loudest voice in the 1920s for a unified and independent air force. It is possible to argue that given Mitchell's personality, his difficulties in getting along with superiors to a point of insubordination, and his self-serving testimony, speeches, and writings, he should not have been so elevated. He did, however, sink the *Ostfriesland* and constantly promote the air service. Billy Mitchell was a balancing act, not really understood by those of the Department of Air Force establishment, but needed by them. It is quite possible that Mitchell would not have gotten along with them any better than he did with the establishment of the 1920s.

At the international airport in Billy Mitchell's hometown of Milwaukee there is an impressive Mitchell Gallery of Flight Air Museum. A great deal of time has been spent to chronicle Mitchell's successes and the development of airpower and flight. A photo shows Mitchell standing in front of his airplane, the Osprey, smiling, holding his hallmark walking stick, wearing a leather flying jacket. One does not see the lonely lad asking for a baked possum, a humiliated Mitchell being expelled from his office near the Western Front, a Mitchell embittered as magazines turned down his articles because they had become tiresome, redundant, angry. He was known as the "Stormy Petrel of the Air"—petrel, a bird in flight. He was that, and he soared with eagles when the *Ostfriesland* went down, but as he grew older it seemed almost impossible for him to lift from the earth for even a short time. In the end, though, one could well say, what a flight it had all been for Billy Mitchell.

NOTES

1. *The Milwaukee Journal,* 21 February 1936, 1.
2. Ibid.
3. Ibid., 22 February 1936, 1.
4. Ibid.
5. Clipping found in the George Hardie Collection, Golda Meir Memorial Library Archives, the University of Wisconsin, Milwaukee, Carton 10, Folder 4. (Hereafter cited as the Hardie Collection.)

6. Information provided by the courtesy of Phyllis F. Aurand, Thomas Balch Library, Leesburg, VA.

7. Gerard M. Devlin, *Paratrooper* (New York: St. Martin's Press, 1979), chapters 2 and 3. Devlin's work, which is authoritative, does not mention Mitchell in connection with the Test Platoon, but does cite his ideas in 1918 in chapter 1.

8. *Chicago Tribune,* 9 October 1940, in the William Mitchell Papers, Library of Congress, Washington, DC, Carton 53. (Hereafter cited as the Mitchell Papers.)

9. James J. Cooke, "Blitzkrieg," in David T. Zabecki (ed.), *World War II in Europe,* Vol. 2 (New York: Garland, 1999), 1175–1177.

10. See, for example, "A Vindication of General Mitchell," *National Aeronautics* (September 1940), 18–19; and Richard L. Harkness, "Vindication of a Valiant Veteran," *Everybody's Weekly* (7 September 1941), *The Philadelphia Inquirer,* 2.

11. *Milwaukee Sentinel,* 24 December 1941, 1.

12. *The State,* 16 June 1942, found in clipping files in the Mitchell Papers, Carton 53. Also see Walter Lippman, "The Whale, the Elephant and the Eagle," *Washington Post,* 14 October 1940. Lippman was one of the loudest voices extolling Mitchell's various warnings and "prophecies."

13. *New York Journal American,* 1 June 1942, found in clipping files in the Mitchell Papers, Carton 53.

14. William F. Trimble, *Admiral William A. Moffett: Architect of Naval Aviation* (Washington: Smithsonian Institute Press, 1994), 90–91.

15. Program and Menu, The Explorers Club, Thirtieth Annual Dinner, 6 January 1934, Mitchell Papers, Carton 50.

16. *Manitowoc Herald Times,* Manitowoc, WI, 15 December 1941, in the Hardie Collection, Carton 10, Folder 4.

17. Information provided by the Alumnae and Alumni Society of Vassar College, Poughkeepsie, NY.

18. *San Francisco Examiner,* 19 December 1941, Clippings File, Hardie Collection, Carton 10, Folder 4.

19. *Milwaukee Sentinel,* 28 October 1942, Clippings File, Hardie Collection, Carton 10, Folder 4.

20. Clippings from the Mitchell Papers, Carton 19.

BIBLIOGRAPHY

Adams, R. J. Q. *The Great War, 1914–1918*. College Station: Texas A&M Press, 1990.

Air Force Academy Library, The Mason Patrick Diaries, 1917–1919.

The Air Force Historical Agency Archives, Maxwell Air Force Base, Alabama.

Alaska Communications Systems, *49th Anniversary, 1900–1949*. Alaska: Alaska Communications Systems Command, 1949.

Arnold, Henry H. "The Cavalry-Air Corps Team," *The Cavalry Journal* 37 (January 1928).

———. *Global Mission*. New York: Harper and Bros., 1949.

"Aviation and Colonel Mitchell," *The Independent* (September 1925).

"The Baker Committee Reports…," *National Aeronautic Magazine* (August 1934).

Baker, Newton B. *Frontiers of Freedom*. New York: Doubleday, Doran, and Co., 1931.

Bayrd, Still. *Milwaukee: The History of a City*. Madison: State Historical Society of Wisconsin, 1948.

Bell, William Gardiner. *Commanding Generals and Chiefs of Staff, 1775–1983*. Washington: Center of Military History, 1983.

———. *Secretaries of War and Secretaries of the Army*. Washington: Center of Military History, 1919.

Braim, Paul. *The Test of Battle: The American Expeditionary Force in the Meuse-Argonne Campaign*. Newark: University of Delaware Press, 1987.

Chandler, Charles Def., and Frank Lahm. *How Our Army Got Wings*. New York: Ronald Press, 1943.

Chief of the Air Service, AEF. *Final Report of the Chief of the Air Service, AEF to Commander-in-Chief*. Washington: Government Printing Office, 1921.

Coffman, Edward. *The War to End All Wars*. Madison: University of Wisconsin Press, 1986.

Cooke, James J. *Pershing and His Generals: Command and Staff in the AEF*. Westport, CT: Praeger, 1997.

———. *The Rainbow Division in the Great War, 1917–1919*. Westport, CT: Praeger, 1994.

———. *The U.S. Air Service in the Great War*. Westport, CT: Praeger, 1996.

Davis, Burke. *The Billy Mitchell Affair*. New York: Random House, 1967.

Dawes, Charles G. *A Journal of the Great War*. 2 vols. New York: Houghton-Mifflin, 1921.

Dobson, Gwen (ed.). *Middleburg and Nearby*. Private Printing, Fauquier County Historical Society, 1988.

Driggs, Laurence L. "The Aviators' Rebellion," *Outlook*, Pt. I, October 14, 1925; Pt. II, October 21, 1925.

Drum, Hugh A. "Trying Our Wings to the Ground: The Argument Against an Independent Air Service," *The Independent* (March 1925).

Fausold, Martin L. *James W. Wadsworth, Jr: The Gentleman from New York*. Syracuse: Syracuse University Press, 1975.

Fickett, Edward M. "A Study of the Relationship Between Cavalry and the Air Service in Reconnaissance," *The Cavalry Journal* (October 1923).

Fischer, William Edward, Jr. *The Development of Night Aviation to 1919*. Maxwell AFB: Air University Press, 1998.

Foulois, Benjamin, and C. V. Glines. *From the Wright Brothers to Astronauts*. New York: McGraw-Hill, 1968.

Frank, Sam H. "Organizing the U.S. Air Service, Part 3: Training Activities in the United States," *Cross and Cockade Journal* 6 (Winter 1965).

———. "Air Service Combat Operations, Part 5: The Toul Sector Operations," *Cross and Cockade Journal* 7 (Summer 1966).

———. "Air Service Combat Operations, Part 6: Operations in the Baccarat, St. Di, Vesl, River Sector," *Cross and Cockade Journal* 7 (Autumn 1966).

———. "Air Service Combat Operations, Part 7: Operations in the St. Mihiel Salient," *Cross and Cockade Journal* 8 (Spring 1967).

———. "Air Service Combat Operations, Part 9: Operations in the Meuse-Argonne Campaign," *Cross and Cockade Journal* 8 (Summer 1967).

Gauvreau, Emile, and Lester Cohen. *Billy Mitchell: Founder of Our Air Force and Prophet Without Honor*. New York: E. P. Dutton, 1942.

"General Mitchell on Aviation," *Aerial Age Weekly* 12 (February 14, 1921).

Greenwald, Byron E. "An Early Answer to Airpower: Anti-Aircraft in the AEF," *Over the Front* 9 (Spring 1994).

Greer, Thomas H. "Air Arm Doctrinal Roots, 1917–1918, *Military Affairs* 20, 4 (Winter 1956).

Gregory, John B. *History of Milwaukee, Wisconsin*. Vol. 3. Milwaukee: S. J. Clarke, 1931.

Guidoni, A. "A New System of Bombing Tests," *Aviation* (October 31, 1931).

Hall, R. Cargill. *Case Studies in Strategic Bombardment*. Washington: Government Printing Office, 1998.

Hallas, James H. *Squandered Victory: The American First Army at St. Mihiel*. Westport, CT: Praeger, 1995.

Hallion, Richard P. *Rise of Fighter Aircraft, 1914–1918*. Baltimore: Nautical and Aviation Publishing, 1984.

Hanson, Arlen J. *Gentlemen Volunteers*. New York: Arcade, 1996.

Harbord, James G. *The American Army in France, 1917–1919*. Boston: Little, Brown, and Co., 1936.

Hartney, Harold. *Up and At 'Em*. New York: Stackpole, 1940.

Hennessy, Juliette A. *The United States Army Air Arm, April 1861 to April 1917*. Washington: Office of Air Force History, 1985.

Hicks, Cuthbert, "America's Air Tangle," *Outlook* (December 20, 1920).

Howard, Louis Conrad. *History of Milwaukee County*. 3 vols. New York: American Biographical Publishing, 1931.

Howells, John, and Marvin Skelton. "Creating General Pershing's War-Time Air Service," Part 2, *Over the Front* 14, 2 (Summer 1999).

Hurley, Alfred H. *Billy Mitchell: Crusader for Air Power*. Bloomington: Indiana University Press, 1975.

Jabolinsky, David (ed). *Roots of Strategy, Book 4, Mahan, Corbett, Douhet, Mitchell*. Mechanicsburg, PA: Stackpole, 1999.

James, D. Clayton. *The Years of MacArthur, Vol. 1, 1880–1941*. New York: Houghton-Mifflin Co., 1970.

Kelly, E. M. "The Man Who Wouldn't Shut Up," *Collier's* 76 (December 12, 1925).

Kennedy, David M. *Over Here: The First World War and American Society*. New York: Oxford University Press, 1980.

Kennett, Lee. *The First Air War, 1914–1918*. New York: Free Press, 1991.

Kroll, H. P. *Kelly Field in the Great War*. San Antonio, TX: San Antonio Printing Company, 1942.

Layman, R. D. *Naval Aviation in the First World War*. Annapolis: Naval Institute Press, 1996.

Lebow, Eilen F. *A Grandstand Seat: The American Balloon Service in World War I*. Westport, CT: Praeger, 1998.

Levine, Issac D. *Mitchell: Pioneer of Air Power*. New York: Duell, Sloan and Pearce, 1958.

Library of Congress: The Aero Club of Washington, DC, Records and Papers; Personal Papers of Henry A. Arnold, Ira W. Eaker, Benjamin Foulois, William Freeland Fullam, James G. Harbord, John L. Hines, Peyton C. March, William D. Mitchell, John J. Pershing, Edward Rickenbacker, Carl Spaatz, and Charles P. Summerall; John Weeks Diaries.

Liggett, Hunter. *Ten Years Ago in France*. New York: Dodd, Meade, and Co., 1928.

———. *Commanding an American Army*. Boston: Houghton-Mifflin, 1925.

McCoy, Donald R. *Calvin Coolidge: The Quiet President*. New York: Macmillan Co., 1967.

Malone, Dumas (ed). *Dictionary of American Biography*. Vol. 19. New York: Charles Scribner's Sons, 1936.

March, Peyton Conway. *The Nation at War*. Garden City, NJ: Doubleday, Doran, and Co., 1932.

Marquis, Albert Nelson (ed). *Who's Who in America*. Chicago: The A. N. Marquis Company, various volumes, 1930s.

Mauer, Mauer (ed). *The U.S. Air Service in World War I*. 4 vols. Washington: Office of Air Force History, 1978–1979.

Mitchell, Ruth. *My Brother Bill: The Life of General "Billy" Mitchell*. New York: Harcourt Brace, 1953.

Mitchell, William. "Aeronautical Era," *Saturday Evening Post* (December 20, 1924.)

———. "Aircraft Dominate Seacraft," *Saturday Evening Post* (January 24, 1925).

———. "Airplanes in National Defense," *Annals of the American Academy of Political and Social Science* 131 (May 1927).

———. "Air Service at the Argonne-Meuse," *World's Work* 38 (September 1919).

―――. "Air Service at St. Mihiel." *World's Work* 38 (August 1919).

―――. "American in the Air," *National Geographic* 39 (March 1929).

―――. "American Leadership in the Air," *Saturday Evening Post* (January 10, 1925).

―――. "Anecdotes of the Air," *Woman's Home Companion* 58 (February 1931).

―――. "The Automobile of the Air," *Woman's Home Companion* 59 (May 1932).

―――. "Building a Futile Navy," *The Atlantic Monthly* (September 28, 1928).

―――. "Civil and Commercial Aviation," *Saturday Evening Post* (February 7, 1925).

―――. "A Glance at World Aeronautics," *Saturday Evening Post* (April 19, 1930).

―――. "Has the Airplane Made the Battleship Obsolete?" *World's Work* (April 1921).

―――. "How Should We Organize Our National Airpower?" *Saturday Evening Post* (March 14, 1925).

―――. "Leaves from My War Diary," *Liberty* (March 31, April 7, April 14, April 21, April 28, May 5, and May 12,1928). [These articles are heavily edited and should be consulted with caution.]

―――. "Look Out Below," *Collier's* 81 (April 21, 1928).

―――. *Memoirs of World War I: From Start to Finish of Our Greatest War*. New York: Random House, 1960.

―――. "Next War in the Air," *Popular Mechanics*. 63 (February 1935).

―――. *Our Air Force: The Key to National Defense*. New York: Dutton, 1921.

―――. "Some Considerations Regarding a Limitation of Armaments," *Annals of the American Academy of Political Social Sciences* 120 (July 1925).

―――. "The Sporting Side of Aviation." *Saturday Evening Post* 26 (April 1930).

―――. "Tiger Hunting in India," *National Geographic* 46 (November 1924).

―――. *Winged Defense*. New York: G. P. Putnam, 1925.

―――. "The World's Largest Airship," *Woman's Home Companion* 58 (November 1931).

―――, et al. *Report of Inspection Trip to France, Germany, Holland, and England, Made During the Winter of 1921–1922*. Washington: Government Printing Office, 1923.

Moore, Samuel Taylor. "Airplanes Land in the Field of Politics," *The Independent* (December 1925).

Morrow, John H., Jr. *The Great War in the Air: Military Aviation from 1909 to 1921*. Washington: Smithsonian Institution Press, 1993.

Mowry George E. (ed.). *The Twenties: Fords, Flappers, and Fanatics*. Englewood Cliffs, NJ: Prentice-Hall, 1963.

National Archives, Archives II, College Park, MD. Records Group 153, Records of the Judge Advocate General (Army), General Courts-Martial, William Mitchell Case Records, 1925, Case Number 168771.

National Archives, Washington, DC. Gorrell's History of the American Expeditionary Air Service, 1917–1919.

―――. Records Group 18, U.S. Air Forces; Records Group 120, Records of the AEF; Records Group 407, Adjutant General Files.

Nofi, Albert A. *The Spanish-American War, 1898*. Conshohocken, PA: Combined Books, 1996.

Parton, James. *Air Force Spoken Here: General Ira Eaker and the Command of the Air*. Bethesda, MD: Adler & Adler, 1986.

Patrick, Mason. "Cost of Our War Time Aircraft," *Current History Magazine* (February 1923).

———. *The United States in the Air*. New York: Doubleday, 1928.

Pershing, John J. *My Experiences in the World War*. 2 vols. Blue Ridge Summit, PA: Tab Books Reprint, 1989.

Pogue, Forrest C. *George C. Marshall*. New York: Viking, Vol. 1, 1963; Vol. 2, 1966.

Prange, Gordon W. *At Dawn We Slept: The Untold Story of Pearl Harbor*. New York: McGraw Hill, 1981.

"Report of General William L. Kenly, The Director of Military Aeronautics," *Air Power* (December 1918).

Rose, Elihu. "The Court-Martial of Billy Mitchell." *Military History Quarterly* (Spring 1996).

Ruffin, Steven A. "Flying in the Great War: RX for Misery," *Over the Front* 14, 2 (Summer 1999).

Scheel, Eugene M. *The History of Middleburg and Vicinity*. Warrenton: Fauquier County Historical Society, 1988.

de Seversky, Alexander. *Victory Through Air Power*. New York: Simon and Schuster, 1942.

Sherman, W. C. "Changes in Air Tactics Since the World War," *U.S. Air Service* (February 1925).

Shiner, John F. *Foulois and the U.S. Army Air Corps, 1931–1935*. Washington: Office of Air Force History, 1983.

Smythe, Donald. *Pershing: General of the Armies*. Bloomington: Indiana University Press, 1986.

Sweetser, Arthur. *The American Air Service*. New York: D. Appleton and Co., 1919.

Thayer, Lucien H. *America's First Eagles*. San Jose, CA: Bender, 1983.

Toulmin, H. A. *Air Service: American Expeditionary Force, 1918*. New York: D. Van Nostrand, 1928.

Trimble, William F. *Admiral William A. Moffett: Architect of Naval Aviation*. Washington: Smithsonian Institution Press, 1994.

United States Army Military History Institute Archives, Carlisle Barracks, PA: Personal Papers of Hugh A. Drum, Ira Eaker, Dennis Nolan, C. H. M. Roberts, and George O. Squier.

University of Wisconsin, Milwaukee. The Golda Mier Memorial Library Archives: The George Hardie Collection; The Mitchell Family Papers.

U.S. Army, Center of Military History. *The United States Army in the World War, 1917–1919*. 12 vols. Washington: Government Printing Office, 1989.

U.S. Congress of the United States. *Biographical Directory of the United States Congress*. Washington: Government Printing Office, 1989.

Vandiver, Frank. *Black Jack: The Life and Times of John J. Pershing*. College Station: Texas A&M Press, 1977.

"Verbatim Report of the Morrow Commission Inquiry," *Army and Navy Journal*, Section II (September 26, 1925).

Waterhouse, Helen S. "D'Artagnan of the Air," *Christian Science Monitor* (October 9, 1935).

Werrell, Kenneth P. *Archie, Flak, AAA, and Sam*. Maxwell AFB: Air University Press, 1988.

"What General Mitchell Claimed," *Aviation* (August 1, 1921).

Williams, George Bruce. *History of Milwaukee, City and County*. Vol. 1. Milwaukee: S. J. Clarke, 1931.
Williams, George K. *Biplanes and Bombsights: British Bombings in World War I*. Maxwell AFB: Air University Press, 1999.
The World War One Diary of Col. Frank P. Lahm. Maxwell AFB: Historical Research Division, 1970.
Writers Program, WPA, State of Wisconsin. *Wisconsin: A Guide to the Badger State*. New York: Duell, Sloan, and Pearce, 1941.
Zabecki, David T. (ed.). *World War II: An Encyclopedia*. New York: Garland, 1999.

NEWSPAPERS AND JOURNALS

Aerial Age, 1920–1925
Aviation, 1920–1925
Baltimore Sun, 1920–1925
Detroit Free Press, 1923
Milwaukee Journal, 1920–1936
Milwaukee Sentinel, 1877, 1882
New York Herald, 1920–1925
New York Times, 1919–1936
San Francisco Chronicle, 1920–1925
Washington (D.C.) Herald, 1920–1925
Washington (D.C.) Post, 1920–1926

INDEX

ABOUT THE BOOK

This compelling chronicle of a controversial figure—a man who could be charming, fanatical, arrogant, and confrontational—places Billy Mitchell in the context of the great debates over U.S. airpower between the world wars. Mitchell demonstrated during World War I that massive airpower could decisively affect combat operations on the ground, and he argued vehemently to anyone who would listen that airpower would be the decisive factor in the next war—a war that he was certain would be fought with Japan. But his brilliance was often overshadowed by his personal failings: typically, he alienated those in power who could act on his ideas.

In a highly publicized trial, Mitchell was court-martialed and found guilty, ostensibly for openly attacking the Navy and the War Department over the fatal crash of the airship *Shenandoah*, but primarily for making public his warnings about U.S. weaknesses in the air. Although the air attack on Pearl Harbor made Mitchell look to some like a prophet martyred for his integrity, Cooke revises that portrait to reveal a character fatally flawed by consuming ambition and a man who was a victim only of circumstances of his own creation.

James J. Cooke is professor emeritus of history at the University of Mississippi. His publications include *All-Americans at War, Pershing and His Generals*, and *The U.S. Air Service in the Great War*.